112 MERCER STREET

ALSO BY BURTON FELDMAN

The Nobel Prize:
A History of Genius, Controversy, and Prestige

112 MERCER STREET

*Einstein, Russell, Gödel, Pauli,
and the End of Innocence in Science*

Burton Feldman

Edited and completed by
Katherine Williams

Arcade Publishing • New York

S

Copyright © 2007 by Burton Feldman and Katherine Williams

FIRST EDITION

Library of Congress Cataloging-in-Publication Data

Feldman, Burton.
  112 Mercer street : Einstein, Russell, Gödel, Pauli, and the end of innocence in science / Burton Feldman edited and completed by Katherine Williams.
  —1st ed.
     p. cm.
  Includes bibliographical references and index.
  ISBN 978-1-55970-704-6 (alk. paper)
  1. Scientists—History. 2. Scientists—Philosophy. 3. Science—History.
4. Einstein, Albert, 1879–1955. 5. Russell, Bertrand, 1872–1970. 6. Gödel, Kurt. 7. Pauli, Wolfgang, 1869–1955 I. Williams, Katherine S. II. Title. III. Title: One hundred and twelve Mercer street.

  Q141.F345 2007
  509.2'2—dc22   2007001194

Published in the United States by Arcade Publishing, Inc., New York
Distributed by Hachette Book Group USA

Visit our Web site at www.arcadepub.com

10 9 8 7 6 5 4 3 2 1

Designed by API

EB

PRINTED IN THE UNITED STATES OF AMERICA

To Peggy Feldman

CONTENTS

PART 3: THE UNIVERSE

PART 4: BEYOND PATHOS: OPPENHEIMER, HEISENBERG, AND THE WAR

INTRODUCTION

GIVEN THE DAZZLING HISTORY of twentieth-century scientific discovery, the story of Albert Einstein, Bertrand Russell, Kurt Gödel, and Wolfgang Pauli in conversation might seem trivial. Nothing really emerged from their meetings, as far as we can tell — neither momentous discovery nor earth-shattering weaponry.

Yet their confluence in the small town of Princeton, New Jersey, during the winter of 1943–44 is fascinating to ponder. They were giants divided by a generation. Einstein was schooled in classical physics; Russell began his career at the very dawn of mathematical logic. By the 1920s, when Pauli and Gödel came of age, certainties had dissolved. For Pauli, especially, the new quantum physics opened up endless possibilities for insight and production.

In this generational battle, Einstein fought desperately against the quantum mechanics so ably theorized by Pauli and his quantum brethren. Unable to refute the new physics, Einstein retreated into what now seems a doomed search for a unified theory. To his contemporaries, it seemed daft, even obstructionist. His failed search illustrates the pathos of science. For scientists, as for athletes, age does not correlate with creativity. The poet may begin a long career with an apprenticeship of pastoral poetry, progressing finally to the noble epic, a trajectory known as Virgil's wheel. But science and mathematics are notoriously the provinces of youth.

As Einstein fought his battle in the world of physics, history was busy transforming how physics worked. World War II turned theoretical physics into a race for the ultimate weapon. Robert Oppenheimer, once a student of Pauli, headed the project in America. Werner Heisenberg, Pauli's friend and partner in the founding of quantum mechanics, was integral to Germany's effort. Inevitably, the bomb conferred on physics both prestige and peril. It took a soldier to coin the phrase "military-industrial complex." Well before Eisenhower's words of warning, physicists understood their plight. The war, said the scientist and novelist C. P. Snow in a speech on morality and science, turned physicists into "soldiers-not-in-uniform."

The pretext for this book is the gathering of four great minds in an academic backwater during a war that would change science forever. "I used to go to [Einstein's] house once a week to discuss with him and Gödel and Pauli," wrote Russell in his *Autobiography*. That these meetings occurred is generally accepted by the biographers of Einstein, Russell, and Pauli. However, as with all things human, there is disagreement, and to ignore contrary evidence would be unseemly, not to mention unscholarly. Thus, against Russell's recollection of those meetings at 112 Mercer Street, we must cast some doubt — that is, uncertainty of a mundane, rather than quantum, sort. Russell himself, speaking on the BBC in 1965, reported that in Princeton, Einstein "arranged to have a little meeting at his house once a week at which there would be some one or two eminent physicists and myself." No mention was made of Gödel. Only in his *Autobiography* does Russell assert that he, Einstein, Pauli, *and* Gödel met regularly.

In 1971, Kurt Gödel learned of Russell's claim and drafted (but never sent) a rebuttal of sorts to a friend: "[t]he passage gives the wrong impression that I had many discussions with Russell, which was by no means the case (I remember only one). . . ." More pointed was Gödel's acerbic retort to Russell's assertion in the same passage that "Gödel turned out to be an unadulterated Platonist. . . ." Gödel responded, "[My] platonism is no more 'unadulterated' than Russell's own in 1921." As we shall see, Russell and Gödel enjoyed a tangled relationship — one that might have encouraged exaggeration about the question of their meeting from either man. It is possible, even likely, that Gödel turned up at Einstein's rather less than Russell's remark suggests, and equally possible that he showed up more than once.

Who was in attendance and when will forever be a matter of conjecture, as must our thoughts on what might have been said, beyond Russell's few (easily assailable) memories. More fascinating is the context — cultural, theoretical, biographical, and historical — of those meetings. The crosscurrents that made up this context are the subject of this book.

Finally, as must be obvious, the perspective of this book is not that of a scientist. Some might argue that the history of science is better left to scientists. For better or for worse, however, nonscientists are enthralled by that richly creative, densely coded world — beyond our grasp as laypersons, yet so enveloping of our lives.

To elucidate major characters and relevant concepts, we provide below brief biographical sketches and a glossary of terms. Terminology in boldface is cross-referenced in the Glossary.

ALBERT EINSTEIN (1879–1955)

With the publication of his theory of **general relativity**, Einstein, already well known among physicists, became world famous. No more ubiquitous a face has ever represented science in the popular imagination. Almost instantly, Einstein turned his fame into a platform for political and humanitarian causes: internationalism, support for Israel, antifascism, civil rights, socialism. At the end of his life, with Russell, he made a final plea for world peace.

Best known among Einstein's works are the two relativity theories: the **special theory of relativity** (1905), which introduced the notion of spacetime, and the **general theory of relativity** (1916), which explained gravitation. In his "miracle year" of 1905, Einstein wrote a total of four papers, including that proposing special relativity. Ironically, it was not for either theory of relativity, but for the first of his 1905 papers that Einstein won the 1922 Nobel Prize in Physics. That paper, on the photoelectric effect, was an inaugural step in the development of **quantum mechanics**, much to Einstein's chagrin. He spent most of his later years in a failed search for a **unified theory** based not on quantum mechanics, but on relativity.

BERTRAND RUSSELL (1872–1970)

Best known for his political activism, Bertrand Russell played a major role in the development of twentieth-century analytical philosophy. At Cambridge, he majored in mathematics and scored well enough on the feared "tripos" exams to be ranked among the first division "wranglers." That Russell was able to bridge such utterly disparate worlds speaks as much to the restlessness of his intellect as to its power.

Russell's startling discovery in 1901 of a paradox that would bear his name, **Russell's paradox** (the conundrum posed by imagining a set of all sets that are not members of themselves), launched him into a decade's worth of work. In 1903, he published *The Principles of Mathematics*, precursor to the much longer *Principia Mathematica*, written with Alfred North Whitehead and published in three parts from 1910 to 1913. This latter work broke philosophical ground by promoting mathematical logic, or **logi-**

cism, and by introducing a theory of types and a powerful notation system. For the first time, mathematically based logic made its way into the mainstream of philosophy, long dominated by metaphysics and epistemology.

Political activism characterized the latter decades of Russell's life. Few left-wing causes escaped his energetic support: pacifism, Bolshevism, antifascism, anti-Stalinism, nuclear disarmament, decolonization, the International War Crimes Tribunal. On the brink of war over the Cuban missile crisis, Kennedy and Khrushchev each received a telegram. (Kennedy's reaction, in private, was to call Russell a "son of a bitch.") Even into his nineties, Russell did not flag. A letter transcribed and mailed on February 3, 1970, a day after Russell died, sent his greetings to representatives of South Vietnam's Provisional Revolutionary Government. A lifetime of writing on philosophy and politics won Russell the 1950 Nobel Prize in Literature.

Kurt Gödel (1906–1978)

As a logician, Gödel has no modern rival. Not since Aristotle, some say, have one man's theories so utterly transformed the field of logic. He was a quiet, remote, unworldly mathematician, given to eccentricities and belief in ghosts. With Einstein, he would make his daily trek to the Institute for Advanced Study in Princeton. Their friendship was famous. In Gödel, Einstein found his intellectual equal.

As a graduate student at the University of Vienna, Gödel sat in on meetings of the Vienna Circle, a highly influential group of philosophers dedicated to empiricism (the idea that knowledge is derived through experience) and logical analysis (held to be the proper method for solving problems in philosophy). Yet in 1931, Gödel proposed his **incompleteness theorems**, ending once and for all attempts to find a complete and consistent set of axioms for all of mathematics. Russell's logicism and the Vienna Circle's logical analysis were dealt a fatal blow, insofar as they sought complete consistency within a system of logic.

Although Gödel contributed to mathematical and logical thought throughout his life, the incompleteness theorems and their metalogical system of notation known as **Gödel numbering** were his crowning achievements. Gödel left Vienna for Princeton in 1940 and never returned, gaining a professorship at the Institute for Advanced Studies in 1946 and publishing a tantalizing demonstration of time travel based on general relativity in 1949. He died in 1978 of starvation, so overcome by paranoia that he believed his food was being poisoned.

Wolfgang Pauli (1900–1957)

Few outside the world of physics know of Wolfgang Pauli. His fame rests on two discoveries: the **exclusion principle** and the **neutrino**. The former is an esoteric explanation of how electrons behave in the atom. The latter is a bit more fathomable among laypeople — the neutrino being first of many particles found to exist within the atom, beyond the electron, proton, and neutron.

Within the world of physics, however, Pauli is legendary. His exclusion principle is indispensable to our understanding of matter. In 1930, he postulated the existence of the neutrino, the confirmation of which came a year before Pauli's death. Equally important, though, was his centrality to quantum theory. In the 1920s, as the world entered a depression and headed toward world war, Pauli helped foment a revolution that overturned classical physics. Among his fellow revolutionaries were Niels Bohr (the "father" of quantum theory), Werner Heisenberg, Erwin Schrödinger (of "Schrödinger's cat" fame), Paul Dirac, Max Born, and Louis de Broglie.

So integral was Pauli to the development of quantum theory and quantum mechanics that he became known as the repository of "conscience" within the movement. Personable, brash, painfully critical, intellectually honest, he was an inveterate collaborator and an indefatigable letter writer. He won the 1945 Nobel Prize in Physics for the exclusion principle. Having left Europe for the duration of the war, he returned to Switzerland in 1946 and headed the physics program at Zurich Polytechnic School (ETH) until his death.

J. Robert Oppenheimer (1904–1967)

Erudite and precocious, Oppenheimer left Harvard in 1925 with a degree in chemistry and sufficient background in physics to join the Cavendish Laboratory in Cambridge. At twenty-three, he was awarded a doctorate from the University of Göttingen, where he worked with such giants in quantum mechanics as Wolfgang Pauli and Max Born. Having gained respect from his European colleagues, he accepted a position at the University of California, Berkeley, in 1929.

Oppenheimer made no great discovery in physics. As an intellectual and administrative leader, he was second to none. Over the next decade, he mentored dozens of graduate students, dabbled in leftist politics, and favorably impressed the great Ernest Lawrence, who ran the Berkeley Radiation Laboratory. As the American bomb effort began taking shape, Oppenheimer rose from adviser to director of Los Alamos, the secret and central laboratory where the greatest minds in physics concocted the atomic bomb. At the first successful test of an atomic weapon, he later said, words from the *Bhagavad Gita* echoed in his mind: "I am become death, the destroyer of worlds."

During and after the war, he was targeted by government security forces for his leftist politics. His support of an international approach to the uses of atomic fission and his opposition to the hydrogen bomb led to investigations by the House Un-American Activities Committee, and in 1954, his security clearance was revoked. He continued to lecture and work on the fringes of power. From 1946 to 1965, he was director of the Institute for Advanced Study in Princeton. He won the Fermi Prize for Physics in 1963. He died in 1967, having never regained his security clearance.

WERNER HEISENBERG (1901–1976)

Unlike his friend Pauli and his mentor Niels Bohr, Heisenberg tended to think in flashes of brilliance. In 1925, he took on the murky question of quantum mechanics. If we cannot see within an atom, he reasoned, let us use what we can observe, namely, how atoms emit and absorb light. From that thought came **quantum mechanics**. To calculate the movements of particles, Heisenberg stumbled into what became **matrix mechanics**. Meanwhile, Erwin Schrödinger countered with a less abstract explanation likening electrons to waves (**wave mechanics**). To reconcile the seemingly bizarre quantum world with the more familiar classical physics, Heisenberg came up with the **uncertainty principle**: It is possible to measure the location of an electron and the momentum of an electron, but never both simultaneously. For his work on quantum mechanics, he was awarded the Nobel Prize in 1932.

Heisenberg was a fervent nationalist. When the Nazis came to power, he continued his atomic work, participating at the highest levels in the German effort to harness fission. He traveled often to conferences and, on one famous occasion, to visit his old friend and teacher Bohr. That conversation became the subject of Michael Frayn's drama *Copenhagen*. At the end of the war, Heisenberg was interned along with nine other German scientists at Farm Hill in England, where from July through December 1945 their conversations were secretly recorded by British agents. From the end of the war until his death, Heisenberg labored to rebuild atomic physics in Germany.

Glossary

Beta-decay: radioactive decay, or the emission of a beta particle or electron from an atom

Black-body radiation: thermal radiation (that is, heat radiating) from a closed system heated to a particular temperature

Born's probability interpretation: a reconciliation of wave mechanics with quantum mechanics that explains waves as containing the probable location of an electron

Bose-Einstein statistics: first proposed by S. N. Bose and championed by Einstein, these statistics explain the behavior of **bosons.**

Bosons: one of two general classes of elementary particles (along with **fermions**) determined by how they spin

Bright-line spectra: the unique spectral lines formed when an element is heated and its atoms emit light; every atom has its own signature of bright lines.

Brownian motion: in fluids, the seemingly random movements of tiny particles as they are struck by the molecules of the fluid; Einstein applied statistical mechanics to explain the movements.

Copenhagen interpretation: under Niels Bohr, the most persuasive, complete framework for understanding quantum mechanics

Double-slit experiment: a method for analyzing light by diffracting light beams through two slits and observing the resulting patterns on a screen

Exclusion principle: Pauli's discovery, expressed as a law, that no two particles occupy the same space at the same time

Fermions: see **Bosons**

Field: the extension of a physical quality throughout space (as the electromagnetic field); in classical or quantum physics, field theory describes the dynamics and effects within the field.

General Relativity: Einstein's generalization of his theory of special relativity to include gravity. It reconceived Newton by showing that apples fall to the ground because the earth's mass curves the adjacent space-time, forcing apples to move in a special way, that is, towards the surface of the earth. It has proven extremely difficult to unify general relativity with quantum mechanics. **String theory** is currently the best hope.

Gödel numbering: the assignment of numbers to mathematical symbols and formulas, to allow for mathematical statements about mathematics (that is, metamathematical statements)

Incompleteness theorems: Gödel's demonstrations that in any formal system, there are statements that are true, but are not provable using the axioms of that system. The paper containing the theorems was entitled "On Formally Undecidable Propositions of *Principia Mathematica* and Related Systems," a reference to the work by Russell and Whitehead.

Induction: reasoning that begins with information from particular instances and leads to general propositions

Locality: in classical physics, the idea that two objects in separate places are independent and cannot interact, a restriction seemingly disproved by recent experiments demonstrating quantum particles to be "entangled" no matter what the distance between them

Logicism: an approach to philosophy in which mathematics is subsumed into logic

Matrix mechanics: the first complete definition of the laws and properties of subatomic particles using matrices to describe their properties

Maxwell's equations: four equations that describe electric and magnetic fields and their interaction with matter

Neutrino: an extremely light particle, the existence of which was hypothesized by Pauli

Paradox: in logic, reasoning that leads to contradiction, revealing false assumptions or faulty processes

Particle physics: the study of the basic subatomic elements and the forces acting upon and among them

Photoelectric effect: the emission of electrons when exposed to electromagnetic radiation and the subject of Einstein's first 1905 paper, in which he proposed that rather than waves, light was made of quanta (later called **photons**)

Photon: the elementary particle that forms light

Planck's constant: the ratio of a **photon**'s energy to its frequency

Quantum mechanics: a theory of subatomic systems that derives information about particles through an application of statistics, conceives of electrons and protons as both particle and wave in their behavior, and acknowledges uncertainty in measuring both movement and position simultaneously

Quantum physics: the entire body of modern, postclassical physics incorporating **quantum mechanics** and treating both large and small scale forces

Russell's paradox: a conundrum discovered by Russell in 1901. The problem occurs when, in attempting to account for and classify all sets, one imagines a "set of all sets that are not members of themselves." A set can be a member of itself: Imagine the set of all objects that are not cars. Since the set of all noncars is not a car, it can be a member of itself. Now, we must move up a rung in abstraction, because set theory conceives of "sets of sets." The set of all sets that are members of themselves is, indeed, possible: e.g., the set of all sets of cars can be a member of itself. But the set of all sets that are *not* members of themselves is a paradox. Is it a member of itself? It must be, since it is a set that includes just that: sets that are not members of themselves; and yet it cannot be, since by being a member of the set of all sets that are not members of themselves, it would become a member of itself.

Special relativity: Einstein's explanation of the relationships among light, energy, and matter, introducing the concept of space-time and defining the speed of light as the one constant not dependent on the observer

Spin: in quantum mechanics, a particle's angular momentum — either half-integer (**fermions**) or integer (**bosons**)

String theory: a theory (as yet unproven) that attempts to unify all of the known forces (weak, strong, electromagnetic, and gravitational). In this theory, matter is made not of particles that behave like waves, but of strings that vibrate.

Strong and weak nuclear forces: within the atom, the strong force holds neutrons and protons in the nucleus; the weak force allows beta decay, or radiation, to occur.

Ultraviolet catastrophe: a false prediction, based on classical physics, that when **black-body radiation** reaches equilibrium, it will emit infinite heat

Uncertainty principle: a cornerstone of **quantum mechanics**; Heisenberg's assertion that it is impossible to measure and determine both the position and the momentum of a particle simultaneously

Unified theory: a theory of everything, that is, all of the elemental forces (weak, strong, electromagnetic, and gravitational)

Wave mechanics: Erwin Schrödinger's explanation of quantum behavior, in which electrons move like waves around the nucleus

Wave/particle duality: in quantum mechanics, the idea that all objects exhibit the properties of both waves and particles

PART 1

THE PATHOS OF SCIENCE

PRINCETON, WINTER 1943–44

IN PRINCETON, DURING THE COLD, wartime winter of 1943–44, four men — scientists and luminaries all, with common interests and uncommon theories — met once a week over the course of several months. These casual meetings took place far from the horrific battlefields of the World War and far from Los Alamos, the (then) secret lair of experimental atomic physicists.

They were extraordinary meetings, although they probably did not contribute to the advancement of the sciences. The four participants were uniquely matched within the adversarial and communal culture of a pure and disinterested science: Albert Einstein (at whose 112 Mercer Street house the men met); Bertrand Russell, the British logician, philosopher, and gadfly; Wolfgang Pauli, the boy wonder of quantum physics, who formulated the "exclusion principle" in 1925 and postulated the existence of the neutrino in 1930; and Kurt Gödel, whose "incompleteness" theory of 1931 shattered the link between logic and mathematics that Russell's monumental work *Principia Mathematica* had attempted to forge.

We know of these meetings only from passing remarks in Russell's *Autobiography:*

> While in Princeton, I came to know Einstein fairly well. I used to go to his house once a week to discuss with him and Gödel and Pauli. These discussions were in some ways disappointing, for, although all three of them were Jews and exiles and, in intention, cosmopolitans, I found that they all had a German bias toward metaphysics, and in spite of our utmost endeavour we never arrived at common premises from which to argue.[1]

Eventually, the conversations seem to have sputtered out. Russell gives us no other details about what was said, and perhaps nothing worth reporting *was* said.

That such exemplars of our scientific age had occasion to chat in the cloistered world of Princeton's Institute for Advanced Study might seem fodder for yet another London stage play set in the world of physics — Heisenberg and Bohr had met in Copenhagen in 1941, and from that meeting, with its uncertainties and relative perspectives, can be seen, in retrospect, angled perspectives into the collision of particles that is war.

Indeed, it was war and Hitler that had brought all four to Princeton. Einstein, Pauli, and Gödel, having fled the chaos of Europe, found refuge at the Institute (Einstein in 1933, Pauli and Gödel in 1940). Russell was in temporary exile from England, yearning to return, but as yet unable, owing to travel restrictions. Solidly entrenched in the small-town atmosphere, the four men spent their days thinking, writing, and, periodically, lecturing, either within the Institute or elsewhere at professional meetings.

For years, Einstein had enjoyed world fame. Princeton was no different. Shop owners hoarded his signed checks, children pleaded for help with homework, strangers approached him on the street or in museums. He and his second wife, Elsa, had moved to the modest white house at 112 Mercer Street in 1935, just a year before

Elsa's death. During Elsa's illness and after her death, Elsa's daughter, Margot, and Helen Dukas, Einstein's secretary, who had been part of the family since 1928, ran the household. They were joined by Maja, Einstein's gifted and beloved sister, in 1939. By 1943, then, Einstein's household consisted of himself and three exceedingly intelligent women. One of the three would surely have served tea to the guests.

The colonial-style house (which Einstein paid for by selling a manuscript to the Morgan Library) was filled with solid German furniture rescued from the couple's apartment in Berlin. Typically Biedermeier, a style associated with the pretentious nineteenth-century German bourgeoisie, the cumbersome, clumsy, outmoded furniture was relegated to the first floor at 112 Mercer — Elsa's domain. His own study, on the upper floor, was furnished in much plainer fashion, with tall bookshelves and a paper-strewn table.[2] It was there that Einstein often entertained colleagues, and there, presumably, that the four men met.

Clearly, the meetings did not make history. But they certainly embodied it. Einstein's special and general theories of relativity had reshaped modern physics; Pauli's exclusion principle helped launch the revolution in quantum physics; Russell's early eminence as a logician resulted in the towering *Principia Mathematica* (written with Alfred Whitehead), which laid out the foundations of symbolic logic; Gödel's incompleteness theorem quashed any hope (including Russell's) of mathematics as a universal, consistent, and complete system.

A more illustrious scientific group of friends probably never gathered, at least not in such a relaxed and intimate setting as Einstein's study. After all, Einstein was Einstein, and Gödel was considered the most important logician since Aristotle. In terms of stature, their only equivalents would be Newton and Leibniz in the seventeenth century, two geniuses who never met. Of course, conferences and congresses were typical meeting places, with their rituals of podium, prepared papers, and hallway talk. But the conversations Russell alludes to would have had nothing in common with academic conferences.

It was an intimate little group. Gödel and Pauli were among Einstein's closest friends in Princeton. Einstein was fond of Russell, who, as the only nonmember of the Institute for Advanced Study, might have seemed a bit of an interloper. Later, Russell would develop a friendship of sorts with Wolfgang Pauli.

It was also a pot simmering with outsized personalities. Pauli, then aged forty-three, was famously intimidating, given his brilliance and rough sarcasm — and he looked the part: florid and barrel shaped, with strangely slanted eyes in a moonface. (George Gamow, the mathematician, once sketched Pauli as a plump devil, an allusion to his reputation as an intense critic, as well as to his devilish wit.) Gödel, aged thirty-seven, a mathematical Platonist happily married to a former cabaret dancer from Vienna, was a notorious recluse, shadowed by bad health and nervous breakdowns. He was thin and intense, his eyes shielded by horn-rim glasses with tinted lenses — he might be staring at you, but it was hard to tell. The seventy-two-year-old Russell, small and bony-featured, was famous for his lightning mind, malicious wit, and boundless energy; he was now the third Earl Russell and one of the world's leading philosophers, but he was also impoverished and had been in America since 1938 trying (vainly) to find a paying job. Then there was Einstein, his hair wild as ever, wearing a sweatshirt with a handy pen clipped to its neckline.

German accents mingled with upper-class British tones: Einstein talked very softly, Russell in a snapping high pitch, Pauli in a growl, and Gödel quietly and precisely. It would have been a smoke-filled room as well: Einstein, Russell, and Pauli were incessant pipe smokers.

One fascinating topic probably never came up. Two of the four scientists dabbled in the realm of the irrational. Pauli the physicist was a scourge of any idea that did not meet the most rigorous, objective standards of proof. But privately he subscribed to Carl Jung's psychology, with its archetypes and archaic myths. Indeed, he had been Jung's patient in the early 1930s, and Jung had mined his client's rich dream life in several lectures and articles, suppressing Pauli's name, of course. Among Jung's writings on Pauli's dreams is the chapter "Individual Dream Symbols in Relation to Alchemy," which alludes to another of Pauli's extra-rational interests: medieval alchemy. Pauli found in alchemy a model for how matter could be joined to spirit — a task he felt urgent for the modern mind. As for Gödel, the astringent logician was highly sympathetic to telepathy, reincarnation, and the existence of ghosts — not despite being a logician, but because of it: He thought such phenomena rationally justifiable. The young Gödel may have attended a séance; certainly he read widely on parapsychology throughout his life.

Had they been aired, Pauli's Jungian affinities and Gödel's excursions into the paranormal would have startled the two elder scientists. Einstein and Russell both adhered to more conventional metaphysical views. Einstein, though close to both Pauli and Gödel, seems to have been unaware of their otherworldly interests. Perhaps it is no wonder, as Russell complained, that the group could not find common premises, at least when it came to philosophizing about science.

AGING GENIUS

Besides tangled friendships and rivalries, the four men who met in Einstein's living room shared something else: the common fate of being past their prime. By 1944 each had already done his important scientific work. Einstein had already spent the last twenty years pursuing his hope of a grand unified theory for physics — a pursuit destined for failure. Russell's creative years in logic lay more than thirty years in the past; now, he was busily churning out popularizing books to earn badly needed money. (The *History of Western Philosophy,* which he was finishing up, surprised him by becoming a runaway best seller the following year.) Pauli's many achievements included his 1925 discovery of the exclusion principle, one of the great clarifying insights in physics, and his daring prediction in 1930 of the neutrino; but there were no more such triumphs ahead. Gödel's startling theorem of 1931 proved that mathematics must remain incomplete — the "most significant mathematical truth of this century," as his honorary degree from Harvard later put it; but by 1944, he had shifted from logic to make a new start in philosophy.

The history of science is a long procession of figures, some famous, many forgotten, who come forward, work their special wonders, make their mark, lose the power of genius, and make their exit. That scientific careers peak early is, of course, a commonplace — some might say a myth. Yet it would seem that those exits are, on the whole, earlier than those of poets and artists. Aging genius has been a topic of much discussion among scientists and their anatomists. Like any myth, that of the coupling of youth and scientific discovery is both exaggerated and compelling. Indeed, those to whom scientific genius is ascribed have done as much as any to propel the myth. G. H. Hardy, a numbers theorist who continued to produce well beyond middle age, nevertheless wrote in his memoir, *A Mathematician's Apology,* "No mathematician should ever allow himself to forget that mathematics, more than any other art or science, is a young man's game." And we have Russell's own account of his dogged work on the *Principia* to remind us that scientific creativity requires immense energy:

> So I persisted, and in the end the work was finished, but my intellect never quite recovered from the strain. I have been ever since definitely less capable of dealing with difficult abstractions than I was before.[3]

Mathematical theorizing seems especially suited to the young, nor is theoretical physics a country for old men. Thus, we have

John Nash at twenty-two (equilibrium and game theory); John von Neumann at twenty (a definition of ordinal numbers); Carl Friedrich Gauss at twenty-one (the fundamental theorem of algebra); Evariste Galois at twenty-one (recognized posthumously for his algebraic theories); Alan Turing at twenty-four (the Turing machine). Arguably, the average age of discovery has risen in recent years. Jordan Ellenberg reminds us that modern mathematics is itself a mature field, requiring years of study well beyond that of an eighteenth or nineteenth or even early twentieth century prodigy.

Still, Gödel was only twenty-five when in 1931 he proved the "incompleteness" of mathematics. And the early days of quantum physics saw a procession of theorists remarkable for their youth as well as their genius: Wolfgang Pauli and Werner Heisenberg were both twenty-five, and Paul Dirac was twenty-four, when they published their landmark contributions. Newton's insights into gravity and optics came when he was twenty-three, "in the prime of my age for invention" or his "annus mirabilis." Einstein published his special theory of relativity at the age of twenty-nine in his own annus mirabilis of 1905. In the last four centuries, only Newton and Einstein among major theorists were able to surpass their earliest work — Newton with his universal law of gravitation in 1686 and Einstein with his general theory of relativity in 1916.

But even Einstein's gifts finally failed. In the late 1920s, while in his late forties — an almost Methuselah-like age for topflight creative work in theoretical physics — Einstein began his exit.

Why such a "running down" of energy happened to him, or happens to other scientific theorists, is a puzzle, as mysterious as the initial outburst of genius that occurs early in such careers. Explanations range from the physiological (a decrease in testosterone, according to the psychologist Satoshi Kanazawa) to the sociological (math and physics reward brash, revolutionary discoveries, and thus brash, youthful thinkers) to the biohistorical (age statistics are affected by life expectancy or the relative "age" of the field).[4] Despite a mountain of studies, we know very little about why such supreme gifts appear or disappear.

"Genius" is a word that only gives a local habitation and a name to an unfathomable phenomenon: Would we call it genius otherwise? Genius eludes definition partly because it points not to a single ability, but to a web of abilities and coincidences that must hang together delicately yet powerfully to work at all. In physics and logic, mathematical prowess is obligatory; but so are audacity and courage, penetrating insight, imaginative vision, tenacity — and luck. And luck must occur not only within the arena of study — for instance, finding the right equations — but in a larger sense as well: In order to ponder gravity, Newton must first have survived the plague; in order to develop his general theory of relativity, Einstein must first have had the leisure and salary offered by the Kaiser Wilhelm Institute of Physics and must first have accepted the post in late 1913, before World War I had barred him from entering Germany. Accident not only historical, but emotional can loose creative genius. Take, for instance, Einstein's separation from his first wife, Mileva, in the year prior to general relativity, and Gödel's serendipitous marriage to a cabaret singer who kept him sane.

At some point, however, one or another of these powers — or the way they join together — changes, and "genius" departs. Perhaps insight fails, or else ambition or energy, or the theorist becomes too cautious, or is surpassed by students (as the great physicist Max Born confessed of Pauli, his onetime assistant), or can't keep up with new ideas. (Paul Ehrenfest, Einstein's cherished friend, committed suicide at fifty, an act Einstein attributed in part to the "difficulty that adaptation to new ideas inevitably imposes on a man of fifty."[5]) Or perhaps the theorist simply finds physics no longer quite so important — as with Newton, who in middle age took up theology.

The "middle-aged" scientist often seems to his or her colleagues to be a flawed version of that younger, more brilliant self.

Those phenomenal gifts are still evident, but something seems "off" or amiss.

Einstein's seeming failure to produce theoretical insights after the mid-1920s is an example. In 1916, when he published his theory of general relativity, he was thirty-eight, still at the height of his powers. His last contributions, on wave theory and quantum statistics, came in 1925. By then, Einstein was launched on his quest for a unified theory — as quixotic a journey as that of any hapless knight. Many critics think that Einstein misjudged the problem of a unified theory as he would never have done before. But even so, something slowed him down. He was stymied at every step. Only his old boldness and tenacity kept him doggedly going, decade after pointless decade. Some crucial power seems to have left that marvelous mind, and Einstein eventually became resigned to it.

Something similar happened to Pauli in physics and to Russell and Gödel in logic. In his thirties, Pauli was already becoming more of a critical than a creative force within physics. In their thirties, Russell and Gödel both began to abandon logic for philosophy.

Why do physicists and mathematicians seem more susceptible to this fading of creative energy and genius than, say, artists and composers and writers? Johann Sebastian Bach, after all, kept building incomparable structures of abstract musical symbols until his death at seventy-five; Yeats wrote great poems into his seventies. Goethe's artistic maturation continued until his death at eighty-three. Picasso worked successfully into his nineties, DeKooning into his eighties, Braque until his death at eighty-one. No such example of late-stage genius and production can be found in the sciences.

This striking difference between the worlds of science and art was noted by the eminent astrophysicist S. Chandrasekhar, who, musing on artistic genius, thought of Beethoven's words: "*Now*, I know how to compose." At forty-seven, Beethoven had already written eight symphonies, five piano concertos, eleven quartets, and twenty-five of his thirty-two piano sonatas.

Scientists do not develop this way, suggests Chandrasekhar. Their genius flowers young and does not "mature."[6] No major scientist has continued to grow through life as did Beethoven, or Shakespeare, or Rembrandt — not to mention Verdi, Titian, Picasso, or Thomas Mann. To paraphrase F. Scott Fitzgerald, there are a great many triumphant third acts in art and literature, but even second acts are rare in science.

Something systemic — peculiar to scientific endeavor — seems to be involved. Genius is an individual matter — one that varies from person to person and is expressible in a myriad of forms. But science is a collective enterprise. It progresses and builds, dependent on a process of incremental contribution. Even revolutions in science — Thomas Kuhn's paradigm shifts — require structures to build upon (or tear down).[7]

Scientists thus often live to see their greatest work superseded, modified, or refuted. None escapes the relentless march: Immortal Newton was displaced by Einstein, and Einstein fully expected that his work would be corrected or surpassed. In 1949, he wrote to a friend:

> You imagine that I look back on my life's work with calm satisfaction. But from nearby it looks quite different. There is not a single concept of which I am convinced that it will stand firm, and I feel uncertain whether I am in general on the right track.[8]

Einstein's humble reflection reveals from within what can be called "the pathos of science." Neither artists nor philosophers are prey to this pathos: Nothing can "improve" upon Socrates' *Oedipus Rex* or Mozart's *Don Giovanni* or Plato's *Republic*, though these works are subject to the vicissitudes of changing tastes and interpretations. Indeed, poets are especially caught up in intimations of immortality, as distinct from mere fame. Milton invokes the "Heavenly Muse" to ensure that his *Paradise Lost* would "soar /

Above th' Aonian mount, while it pursues / Things unattempted yet in prose or rhyme."

Scientists, on the other hand, create only to be superseded. The greatest scientific achievements will be scrutinized and, eventually, proven inadequate. After two thousand years, Euclid's geometry was shown to be limited and was augmented by the non-Euclidean geometry of Carl Friedrich Gauss. Within twenty years, a fellow German, Georg Friedrich Bernhard Riemann, improved on Gauss — and Riemann's geometry helped lead Einstein to his general theory of relativity.

Science is a community of interlocked, perpetual, cumulative effort. No one can be successful except by working within its common premises and rules of procedure and proof — however "revolutionary" the work. But the cumulative nature of science also means that each individual effort will be supplanted. Discoveries keep occurring — and every discovery means that some previous finding becomes modified or discarded. Thus, most productive scientists become half-forgotten figures in the public mind, existing in textbooks as abbreviations, symbols, and identifiers: Boyle's law, the joule, the fermi, Planck's constant. The incessant construction of science provides new and exalted triumphs (Einstein, after all, can build on Newton), but also ensures one's own "defeat" or limitation — sometimes within a few years.

The four men in Einstein's study provided striking examples of this "pathos of science." Here were two aging scientists paired with upstart revisionists: Einstein and Pauli, Russell and Gödel. At stake were none other than the fundamental structures of modern physics and logic.

Wolfgang Pauli was only sixteen when Einstein's general theory of relativity turned physics upside down. Within four short years, Pauli was to write a definitive explanation of relativity for the *Encyclopädie der Mathematischen Wissenschaften* — an account so clear that forty years later, Niels Bohr lauded it as "still one of the most valuable expositions" of Einstein's theory. Five years later, Pauli presented his "exclusion theory," the first in a number of successive discoveries by Pauli and Werner Heisenberg that defined the nascent field of quantum mechanics.

Behind it all was Einstein, who, using Planck's concepts, "launched" quantum physics in 1905. But Einstein's quantum physics was built on classical physics. No "uncertainty" there. The new quantum mechanics, formulated by Bohr, Heisenberg, and Pauli, no longer postulated an objective reality that could be observed and measured. To Einstein's horror, physics had become a matter of statistical laws rather than certainty. "God does not play dice with the world!" he exclaimed.

Pauli and the quantum physicists had triumphed, however. At the Fifth Solvay Conference in 1927, quantum physics took on classicism from the lectern and in the corridors of discussion. Einstein, who did not present a paper, spoke against the new world of physics heralded by Niels Bohr and his young followers, among them Pauli and the German wunderkind Werner Heisenberg. Stubbornly, Einstein held out against the tidal wave of quantum physics. All indeterminacy was temporary, he insisted, a passing stage within the history of physics. Sooner or later, with more knowledge and insight, physicists would be able to lay aside uncertainty. He never gave up that belief.

But quantum physics, argued Bohr and his quantum conscripts, was here to stay. Uncertainty was not an imposition of humanity onto nature, but a fundamental state. However distressed the Solvay participants might have been by Einstein's vehement opposition, the conference shifted the ground so vigorously that, during the three years between the Fifth and the Sixth Solvay Conference, Einstein found himself in a rearguard position. He was by far the most visible and vocal critic of quantum physics. He devoted the remainder of his life to the search for a unified theory, in hopes of proving Pauli and his quantum mechanics wrong. But in the intervening years, Einstein's position had gained no ground.

Pauli and his quantum associates held sway in a world of physics that had passed Einstein by.

Russell and Gödel were also scientific rivals. Russell's pioneering *Principia Mathematica* won him fame as a logician and was the basis of his philosophic authority and later reputation. Written with Alfred Whitehead and published in three volumes beginning in 1910, the *Principia* tackled the entire domain of mathematics. Its purpose was to demonstrate that "all pure mathematics follows from purely logical premises and uses only concepts definable in logical terms."[9] The impulse to subsume mathematics into pure logic (called "logicism") began with Gottfried Leibniz, who, in the seventeenth century, yearned for a universal language based on logic. Not until the late nineteenth century, however, did logicians develop the tools (in the form of definitions and methods) needed to place mathematics more or less within the realm of logic. In 1879, the German logician Gottlob Frege began his life's work on a system to formalize logic and to develop a logical foundation for mathematics. In their *Principia,* Russell and Whitehead solved inconsistencies that Frege and others could not. (Indeed, Russell had to solve his own "Russell's paradox," which demonstrated inconsistencies in Frege's axioms, before he could complete his *Principia.*) In its three lengthy volumes, the *Principia* devised a comprehensive and very usable notation system; by demonstrating the power of logic, it inaugurated the field of metalogic; it placed logicism comfortably within the realm of traditional philosophy and even made it fashionable.

But the underlying premise of the *Principia* — that mathematics was a complete and thus a universal language and logical system — was thoroughly demolished by the upstart Gödel. In 1931, Gödel published his infamous proof known as the "incompleteness theorem." In it, he demonstrates that no mathematical system that depends on axioms can be thought of as complete, for in any such system, some propositions can be neither proved nor disproved. In extinguishing the dream of a consistent mathematical system, Gödel became, in the eyes of many, one of the two most important logicians of the twentieth century — the other being Russell himself.

Pauli and Gödel were simply following in the tracks of Einstein and Russell. Einstein built upon and upended Newton; Russell built upon and upended the pioneering mathematical logician Gottlob Frege. So science marches on.

SCIENCE AND SIN

As Einstein, Russell, Pauli, and Gödel sat talking in Princeton, Robert Oppenheimer and his fellow physicists in Los Alamos were trying to build an atom bomb that could decide the course of World War II. They believed — and it is certainly argued — that Werner Heisenberg and his fellow physicists were doing the same in Germany.

Physics was a small world. Einstein and Pauli had direct links to these projects, both personal and professional. Fearful of a German bomb, Einstein had written to President Roosevelt in 1939, urging him to begin an atomic project. Pauli had been Oppenheimer's teacher in the early 1930s and had been Heisenberg's close friend and collaborator in the 1920s when, together with Niels Bohr, they had built the foundations of quantum mechanics. In this tightly knit world, Einstein and Pauli were aware of what was at stake, if not in detail.

When the Bomb exploded over Hiroshima, science lost its innocence, irrevocably. In return for the ultimate weapon, physicists tasted dizzying political power. Yet that power was poisoned fruit. Indispensable to their government's survival in the atomic age, physicists were enlisted as guardians of the state — guardians who were kept under strictest guard. Their knowledge was dangerous, and their loyalty was constantly questioned. Oppenheimer's security hearing of 1954 is often thought to have inaugurated the era of atomic suspicion. In fact, distrust and surveillance followed almost immediately from Einstein's letter to Roosevelt. By the time Oppenheimer was staking out Los Alamos, spying on physicists was widespread — even obsessive. Most scrutinized of all, perhaps, was Einstein, whose FBI file eventually numbered some fifteen hundred pages. A new age of suspicion was emerging.

The four men came to Princeton from a continent wracked by war. Each made his way to Princeton as much by luck as by any clear design. There they waited as the world changed. And change it did. In 1945, after the Bomb was dropped, Pauli, critical and prescient as always, lamented:

> The Atom Bomb is a very evil thing, also for physics, I think. The politicians, of course, are at a complete loss and talk in a demagogic way of a *secret* which, evidently, does not exist (the true secret is the nature of the nuclear forces). Although most people say that I see ghosts, I am afraid that physics gets more or less subdued by military censorship and that free research, in principle, is gone.[10]

AT HOME IN PRINCETON

If Einstein and his friends spent the war sidelined and isolated, they were not idle. Of the four, Einstein and Russell were the most outwardly political. Both had abandoned their early pacifism in the face of Hitler and the threat of Fascism. Of the four, only Einstein made any direct contribution to the war effort, working in a minor capacity for the Navy, despite his status as a security risk. Pauli offered his services, only to be told by Oppenheimer to "keep those principles of science alive which do not seem immediately relevant to the war."[11] Pauli was too acerbic and independent to fit into regulated teamwork and had no taste for the applied science useful in military research. Gödel was unfit physically and mentally. Russell was trapped in the United States, unable to secure sea passage to England, where he hoped to contribute to the war effort.

Thus it was Princeton, for better or worse. Einstein described the town "as a wonderful bit of earth, and a most amusing ceremonial backwater of tiny demigods on stilts."[12] As for its intellectual life, "[a]part from the handful of really fine scholars, it is a boring and barren society that would soon make you shiver."[13] Einstein shunned the public spotlight as much as possible. Yet he could never escape the notice of town and gown alike. He was besieged by requests for appearances, statements, appeals, signatures, speeches, and interviews. But he avoided making political statements and reserved his energy for Jewish causes.

Generally, he kept his distance from Princeton's colony of German and European exiles. He was always in the public eye, yet he remained apart, even from the nexus of intellectual life: "I live like a bear in my den."[14] From 1938 to 1941, the great German novelist Thomas Mann taught at Princeton and lived only a few blocks away. But the two men saw little of each other. Mann's wife, Katia, thought Einstein an "enormously specialized talent" but "not particularly stimulating" and "not a very impressive person."[15] A certain cultural snobbery seemed to have migrated to the American shores. Then again, the Manns' grand style of living — in a mansion with a staff of servants — did not appeal to Einstein, who lived simply, without fuss. One can scarcely imagine the dandyish Thomas Mann in Einstein's trademark sweatshirt. Einstein's few close friends in Princeton were outsiders and mavericks, like Gödel and Pauli.

Of the four men who met in Einstein's living room, Gödel, Pauli, and Einstein held appointments at the Institute for Advanced Study. Established by Abraham Flexner in 1930, the Institute offered fellowships (without teaching requirements) to eminent scholars in the natural and social sciences and in physics and mathematics. Einstein was Flexner's first recruit — and his fame fortified the Institute's prestige. Gödel joined in 1933, though only as an assistant; Pauli, after a visit in the mid-1930s, arrived in 1940 to take up residence until the end of the war.

Gödel and Pauli were not only Einstein's intellectual peers, but also, thankfully, spoke German. Einstein's spoken English was never strong; he was really at home only in German. In Princeton, his old friendship with Pauli deepened, and a new one with Gödel flowered. Pauli was superbly knowledgeable about relativity theory, and, when they were not arguing about quantum theory, he and Einstein collaborated on a paper. Gödel and Einstein saw each other daily for years; their walks to the Institute gave Einstein an intellectual equal with whom to discuss his unfashionable unified theory, though Gödel remained skeptical. Russell, whom Einstein had met years before, showed up in Princeton at the end of 1943, at loose ends, scheduled for periodic lectures in New York, desperate for a ship to take him back to wartime England.

Four more varied — and difficult — people would be hard to find.

PART 2

FOUR LIVES

EINSTEIN

FOR ALL HIS SEEMING EASE WITH THE WORLD, Einstein was intensely private. Outwardly friendly, he disclosed almost nothing of himself to the world. Little wonder, perhaps, that biographers have seized on his bohemian clothing and disheveled appearance, as if they could be keys to his inner being. In his Berlin years, when he reached home, he took off the professor's obligatory wing collar and frock coat and walked around in bare feet and an old sweater. In Princeton, with a sigh of relief at American informality, he wore sweatshirts and baggy pants in public and, in winter, a seaman's woolen cap pulled down over his ears. Like his forebear Newton, Einstein was utterly indifferent to fashion.

Privacy was not easily attained. In Princeton, his wife, Elsa, and secretary, Helen Dukas, guarded the front door of 112 Mercer Street. Princeton was as self-contained and self-assured as Einstein himself, so his daily walk to his office attracted little attention. On one occasion, however, his morning walk was interrupted by a high school student sufficiently guileless to have inveigled a rare interview for his school newspaper. "My life is a simple thing that would interest no one. It is a known fact that I was born, and that is all that is necessary." Thus did Einstein steer the young journalist away from personal questions.[1]

No one who met Einstein seems to have harbored suspicions that his reticence was a pose. Indeed, he led a very active social life. Yet he seems to have avoided revelation — even self-revelation, having remained, as he remarked in an appreciation of Freud, among those "not-having-been-analyzed." One activity that tends to reveal is teaching. But Einstein preferred not to teach. His appointments at the Prussian Academy in Berlin and at the Institute for Advanced Study relieved him of all teaching duties. "I couldn't resist the temptation of a post in which I would be free from all obligations and be able to indulge wholly in my musings,"[2] he wrote of his appointment in Berlin. No one took a Ph.D. under his direction. He did not cultivate protégés or disciples among students. As he said, he was always a loner.

In his later years, Einstein alluded to the price he paid for his single-minded devotion to science. When his close friend Michele Besso died in 1955, shortly before Einstein himself, he wrote admiringly to Besso's widow of Michele's ability to lead a "harmonious life":

> [W]hat I most admired in Michele as a man was his ability to live many years with his wife, not only in peace but in constant accord, an endeavor in which I have lamentably failed twice.[3]

Einstein had chosen the perfection of work, Besso the perfection of life. Besso's sister, visiting Princeton in 1947, told Einstein that she said she had long wondered why her brother had not made some great discovery in mathematics. Einstein laughed and said, "Michele is a humanist, a universal spirit, too interested in many things to become a monomaniac. Only a monomaniac gets what we commonly refer to as results." Then, according to Besso's sister, "Einstein giggled to himself."[4]

THE CONFIDENCE OF YOUTH

It is thought that scientific genius is best nurtured in households rich in learning and culture. If so, Einstein's family was ideal. His father, Hermann Einstein, was an easygoing and good-natured man, not particularly suited to the business world. After several false starts, he opened an electrochemical works in Munich with his brother Jakob. Einstein's mother, Pauline, was the more cultured and widely read parent. She also played the piano.

Einstein was born in Ulm, a vibrant, highly industrialized city in southern Germany. The Einstein family history is typical of German Jewry. In the sixteenth century, a small Jewish community grew in the small town of Buchau, about forty miles from Ulm, where an abbey afforded protection. To that small town, in 1665, came Baruch Moises Ainstein. Like other Jews, Ainstein became a tradesman (in cloth and horses) and enjoyed relative freedom to practice his faith. For two centuries, the Jews of Buchau lived nestled against the Alps in peace. During the mid-nineteenth century, however, Einstein's family began a slow migration to Ulm, where prosperity born of industry beckoned. Hermann was born in Ulm and to Ulm he returned with Pauline, whom he met in Stuttgart, where he had been sent to school.

In 1880, a year after Albert's birth, the family moved to Munich, convinced by Hermann's brother Jakob that riches were to be made from the generation of electricity. Jakob, a graduate of Stuttgart Polytechnic, persuaded the more cautious Hermann to join him in partnership. Jakob would serve as inventor and technician; Hermann would tend to the business side.

Despite Jakob's talent, or perhaps because of it, the ensuing years were lean. Munich and the surrounding Bavaria, conservative and steeped in tradition, had resisted industrialization. Not so the rest of Germany. Although the Einstein brothers seemed to have gotten in on the ground floor, they faced fierce competition from well-established companies outside Bavaria. Time and again, Jakob's innovations proved too ambitious and Hermann's caution too inhibiting. Later, Maja, Albert's sister, described their father's method: He "had a particularly pronounced way of trying to get to the bottom of something, by examining it from every side, before he could reach a decision."[5] It was a mode better suited to a physicist than to a businessman. Albert's formative years were spent watching his father and uncle struggle through several incorporations and dissolutions. Capital borrowed from family and friends was lost time and again, and, in the end, the brothers went their own ways.

In many ways, it was an ordinary childhood. Yet the young Einstein was anything but ordinary. Even the moment of his birth provided a shock. Pauline, glimpsing her firstborn, saw only his "large and angular" head, flattened at the back. In a few weeks, time, his skull rounded out; still, he was overly plump, to his grandmother's horror: "Much too fat! Much too fat!" Slow to speak, he nevertheless appeared quite self-possessed as an infant, able to amuse himself. His sister's arrival on the scene may have upended his universe. She was clearly useless as a toy. "Where are its wheels?" he asked.[6]

For all that the Einstein family struggled, it remained solidly middle class. The loving parents, culturally Jewish but not observant, seem to have indulged Einstein and his sister, Maja, nurturing their inquisitiveness and encouraging their musical talents. Despite protests (including hurling a chair at a prospective music teacher), Albert took violin lessons and became so proficient that he and his mother played piano duets. He was free to wander about his neighborhood at the tender age of four. He was also free to let loose his temper on Maja. She narrowly missed being hit by a bowling ball and was not so lucky when Albert came after her with a hoe. More than anything, he was free to explore intellectually, especially in the realm of mechanics. With an extended family steeped in technical and business know-how, Einstein found ready answers to his precocious questions. He was always eager to observe: At age five, during an illness, he was given a magnetic compass by his father. The device, meant only as a distraction, fascinated and excited the budding physicist.

He acquitted himself honorably enough in his schoolwork. But he was the odd boy out in elementary school and, later, the Gymnasium, where rote learning and sports-worship ruled. Happily, he could turn to his uncle Jakob about algebra. Later, a young medical student, Max Talmey, became a boarder in the Einstein house. With Talmey, Einstein found an equal with whom to converse about physical science and higher mathematics. Geometry opened up the "sacred . . . book" of Euclidian geometry for Einstein: Its "clarity and certainty made an indescribable impression on me."[7]

Very early on, he demonstrated the stubborn independence so evident throughout his life. He scorned organized sports and such youthful pastimes as playing soldier: "Poor people," he once said as a uniformed parade passed.[8] He resented the "mindless and mechanical method of teaching" favored by German elementary and secondary schools. Compulsory examinations seemed to him appropriate for "a penal institution."[9] Again and again, he turned to his mathematical pursuits, a world apart from school and the society of children.

At age twelve, Einstein, whose upbringing was secular, got religion. His overwhelmingly Catholic public school was required by law to ensure that he received training in his own religion. A distant relative was unearthed to do the honors, his nonobservant parents being unequal to the task. At first fiercely reluctant, he succumbed to the lure of Judaism with all the fervor of a convert. He was swept up by religious zeal, forgoing pork and composing religious songs. But as the time neared for his bar mitzvah, his high-spirited belief vanished. "[W]ith breathless attention," he began reading "popular scientific books" that thoroughly contradicted, in his view, much of the Bible. Later, he remembered his subsequent "orgy of free thinking, coupled with the impression that youth is intentionally being deceived by the state through lies."[10] He never had a bar mitzvah.

Had his father's business not turned sour, Einstein would probably have endured the hated Gymnasium and graduated with his peers. Military service would then have swallowed him up, all the more agonizing for his antipathy toward authority. But at age fifteen, he was able to escape. His parents were now living in Milan, where Hermann and Jakob, with the backing of Pauline's family, had reconstituted their failed electrical lighting company. Albert stayed in Munich to finish Gymnasium. It was not a happy year. He was lonely and missed his family. He spent Christmas alone for the first time. As ill luck would have it, his "home room" teacher, or primary Gymnasium instructor, was the Greek professor. Although Einstein excelled at mathematics and the sciences, Greek eluded him. Some of the blame must surely rest with adolescent arrogance: Of what use to him that florid language with its dual voice and impenetrable verb system? At any rate, in the spring, the Greek professor exploded: "Your mere presence here undermines the class's respect for me," he shouted.[11] On the verge of dismissal from school, already far beyond the Gymnasium mathematics curriculum, and miserable without his family, Albert begged a family friend and doctor for an official letter allowing him to leave school and join his parents.

The conundrum of what to do now that Albert was in Milan occupied the Einsteins for months. Back in the family fold, Albert regained his spirits. But family finances were at low ebb. In Munich, Albert had been eligible for state-funded schooling. Now, he would have to earn his keep at a practical job (a solution he instantly rejected) or find a way to continue his schooling. Albert resisted both, perhaps because he relished the freedom of days without lessons. He fell in with a congenial group of youths with whom he explored Milan. He also helped with the family business. Finally, it was decided that he would apply to the renowned Zurich Polytechnic School (later renamed Eidgenössische Technische Hochschule — the Swiss Federal Institute of Technology — or ETH) to pursue a degree in electrical engineering. Lacking a high school diploma, he boned up for the college entrance exam. Unfortunately, for all his efforts, he failed three parts of the exam: French, chemistry, and biology. As always, he did poorly in subjects other than his passions — math and physics.

Calamity was averted when physics professor Heinrich Weber saw those scores. He invited Einstein to sit in on his lectures. From there, entrance by special fiat was almost inevitable. All that was required was a high school diploma. A year in Aarau, a small town outside Zurich, gave Einstein much more than his high school diploma. Boarding with a high school professor and his family, Einstein fell in love for the first time, played his violin incessantly, and haunted the beautiful countryside. Marie Winteler, the object of his affection, was destined to be supplanted, the victim of Einstein's great charms and also his tendency to withdraw his emotions without warning.

The following year, Einstein entered Zurich Polytechnic School (ETH). There, he met Mileva Marić, the sole female student in the physics class. Marić, an ethnic Serb, was, like Einstein, an outsider. Highly intelligent and resolute — attributes necessary for her to have penetrated such a male domain as ETH — she knew that in Zagreb her chances of pursuing a technical degree were nonexistent. Switzerland, with its tradition of liberal thinking, gave her the chance. Although well matched in their intelligence, temperamentally Marić and Einstein seemed poles apart. Increasingly at ease with himself and the world, Einstein was as outgoing as Mileva was somber. On the surface, at least, he exuded charm and nonchalance in social settings. Still, in later years Einstein described himself as a "loner." Solitude afforded him the space to think. Nor did solitude necessarily mean physical isolation. As a boy he could shut out the noise surrounding him in a crowded room and lose himself in a problem. Scientists would do well to live in lighthouses, he once said (perhaps thinking of Spinoza, who took up lens grinding), alone and apart for the sake of thought.[12]

At ETH, Einstein studied with several physicists who would become lifelong friends: Marcel Grossmann, whose lecture notes helped Einstein pass math exams; Friedrich Adler, a socialist and, like Einstein, devoted admirer of the philosopher Ernst Mach; and Michele Besso, who would later work with Einstein at the Patent Office. Besso was a particularly cherished friend. At ETH, they devoured material not generally taught in classes, including James Maxwell's theories on electromagnetism and Ernst Mach's critique of Newton. It was Besso to whom Einstein turned in May 1905 with his "difficult problem." Out of that conversation came the final step toward the special theory of relativity.

Not alone among ETH students, Einstein rebelled against the ordinary demands of the professors — attending lectures and taking exams were bothersome distractions. He devoted most of his time and energies to extracurricular studies: Mach, Maxwell, Hermann von Helmholtz, Heinrich Hertz, Hendrik Lorentz, Henri Poincaré. His love affair with Mileva blossomed. In their fourth year at ETH, they both took a required final exam. Einstein passed, coming in third among five. Mileva failed one crucial part. Distraught, she nevertheless allowed herself to be encouraged by Einstein and hoped to retake the exam the following year. Their marriage was planned. It would remain so for three long years, all the while actively opposed by Einstein's mother.

Having graduated, Einstein struggled to make a living. He and Mileva tried their hand at tutoring, but existence was meager. Einstein failed to gain a teaching post, alone among his ETH friends to be shut out. Just as he had antagonized his Gymnasium professors, Einstein had managed to alienate almost everyone in authority at ETH. Broke and discouraged, he and Mileva returned to their respective homes. In the meantime, he gained Swiss citizenship and landed two successive temporary teaching jobs. Mileva, at her home in Novi Sad, was pregnant. She would later give birth to a daughter, to be named Lieserl. The infant was probably given up for adoption some months after her birth.

In 1902, Einstein moved to Bern, where at last he landed a solid job at the Patent Office. He and Mileva had been apart for a year. At the end of 1902, Hermann Einstein died of heart failure, having finally given his approval for the marriage. Einstein and Mileva married in January 1903. Too impoverished for a honeymoon, they returned to their apartment in Bern. Little Lieserl remained in Novi Sad, kept secret from their life in Bern. Einstein never set eyes on her.

THE MIRACLE YEAR

Like a general marshaling troops, Einstein pursued his goals singlemindedly and relentlessly. His behavior with his family was an example. Newlywed though he was, he wrote paper after paper, all the while toiling six days a week at the Patent Office. Mileva, meanwhile, having failed her exams again, settled into the job of housewife, scribe, adviser, colleague, and, in 1904, mother to Hans Albert, the first of their two sons.

It had been a dream of his early youth, to fly along a ray of life as if surfing. That image became a thought experiment and led to the fourth of his five "annus mirabilis" papers of 1905. That paper, famously titled "On the Electrodynamics of Moving Bodies," demolishes Newton's absolute time and space. They become relative to the speed of light. What became known as the special theory of relativity remains today one of Einstein's two most celebrated contributions to physics. The second appeared in a paper written soon after: "Is the Inertia of a Body Dependent on Its Energy Content?" The answer to this question emerged in a simple and elegant formula: $e=mc^2$.

Yet of his five "miraculous" papers, Einstein reserved the word "revolutionary" for the first: "On a Heuristic Viewpoint Concerning the Generation and Transformation of Light." It was revolutionary — the more so for its contribution to quantum mechanics, a worldview he never ceased to oppose. It was also Einstein's most profound contribution to atomic structure — apart from the equation that culminated in Hiroshima.

Einstein's "light" article postulates that the wave theory of light, while useful and commonsensical, since it accords with our view of light as continuous, nevertheless does not accord with experimental data. Rather, light is particle, "distributed discontinuously in space." The paper attempts to resolve a contradiction between the two phenomena: gas and electromagnetism. Gas, it had been shown, was made up of discontinuous particulate matter — atoms. Light or electromagnetic processes were supposed to be waves that traveled through a medium called the ether. Eventually, thanks to Einstein, the problem of "ether," and, indeed, the need to conceive of it, disappear. But in the "light" paper, Einstein is more concerned with solving the disconnect between experimental data and the idea of light as a wave. What he demonstrates, though for years physicists resist his conclusions, is that light is particulate. It is "quantum" in nature, discontinuous and finite, just as are the atoms of helium in a balloon.

The second "miracle year" paper, "A New Determination of Molecular Dimensions," was not published until 1906. However, as soon as Einstein completed it, he sent it to the University of Zurich as his doctoral dissertation. That paper and the third, "On the Motion of Small Particles Suspended in Liquids at Rest Required by the Molecular-Kinetic Theory of Heat," take up the molecular realm. The aptly titled dissertation reveals Einstein's early interest in fixing the size, and indeed the reality, of atoms. The third paper dealt with Brownian motion — the movement of particles suspended in liquid — and thermodynamics. These three papers alone would have assured Einstein's fame as a physicist. Indeed, so radical were the concepts of light quanta and relativity that only very slowly did the world of physics begin to take note.

For six long years, Einstein toiled at the Patent Office by day and revolutionized physics by night. He clarified his "relativity principle" and extended his work on Brownian motion, publishing paper after paper. Yet when he applied to the University of Bern, he was rebuffed — he had not submitted a proper thesis. Finally, having resisted the demeaning requirement (he had sent a collection of papers, to no avail), he buckled down and cranked out the requisite work. He was hired immediately, and in the spring of 1908 he began to teach. A full-time appointment at the University of Zurich followed in 1909. By this time, his fame had spread. He began to lecture widely and was invited to speak at the first Solvay Conference in Brussels. In 1911, he accepted an appointment at the University of Prague. He and Mileva now had two sons, Albert and Eduard. The move to Prague was financially rewarding. The family could now afford a maid. But Mileva was increasingly isolated and depressed. Einstein's science came first, and now his science had catapulted him into the world's eye. In 1913, Einstein was offered a prestigious position, requiring no teaching, as a member of the Prussian Academy in Berlin. It was an offer he could not refuse. To Mileva's dismay, the family moved again.

Once they arrived in Berlin, Einstein wished only to work. Mileva protested, but to no avail. Indeed, he had fallen out of love with her years before. Einstein later said that "she was cool and suspicious toward anyone who, in some way or other, came close to me."[13] In 1903, he had faced down his family's angry disapproval of his marriage to Mileva by being "stubborn as a mule."[14] Now, just as stubbornly, he meant to be free.

In truth, Einstein had motives beyond his career when he accepted the Berlin position. On a visit to Berlin in 1912, he had met his cousin Elsa for the first time since childhood, and their acquaintance intensified during a visit the following year. Elsa, who would become his second wife, took to the task of protecting Einstein and, with some limited success, grooming him. The move to Berlin, so distasteful for Mileva, made Einstein's new life with Elsa possible.

Mileva had barely settled into a Berlin apartment with the children when Einstein sent her a "memorandum" setting harsh conditions: She must not expect him to see her while at home, nor expect affection of any kind; she could speak to him only when he asked and must answer at once when spoken to. To add to Mileva's misery, Berlin was now the home of Einstein's mother, Pauline. Even Einstein sympathized with Mileva's fears of Pauline: "She feels persecuted," he wrote to Elsa. "Well, there's some truth in it."[15] Not surprisingly, Mileva very soon returned to Zurich. Michele Besso and other friends tried to calm the waters, but to no avail. Then began long negotiations for a divorce.

Einstein always possessed a sharply realistic sense of what was necessary. Thus, even during World War I, when all around him were diverted by the tragedy of war and when his family life was disintegrating, he was still able to concentrate. After eight years of work on the general theory of relativity, he made one last great dash in the second half of 1915. Without question, he juggled the distractions by compartmentalizing himself. He brought his full attention to his work, while somehow coping with (or evading) his tumultuous life. He waged battles on different fronts, each of which clamored for his presence. But he remained inwardly centered.

His ability to focus, to the exclusion of all distractions, was legendary. In 1911, when Einstein was in Prague, he spent evenings at the salon of Bertha Fanta along with a group of young Jews interested in philosophy and literature. There he met the young Franz Kafka and Kafka's close friend Max Brod, also a novelist. (Brod later secured his place in literary history by refusing Kafka's dying request to destroy the latter's unpublished manuscripts, among them *The Trial, The Castle,* and *Amerika.*) Brod was then at work on a novel about the great Danish astronomer Tycho Brahe, later published as *Tycho Brahes Weg zu Gott* (*The Redemption of Tycho Brahe*). It was an allegory of scientific genius, based loosely on historical fact. Brahe's careful measurements of planetary movement in the two decades before 1600 made it possible for his assistant Johannes Kepler to find the true laws of planetary motion by 1610 — measurements Newton later used in his epochal theory of gravitation. Kepler dared to assert that planets move around the sun in ellipses.

Brod took Einstein as his model for Kepler. In the novel, Kepler's genius relies on his gift of "retaining the essential and rejecting everything else or perfecting it."

> His patience never failed; he held all in readiness for the one mysterious moment, in which from near and far his "laws" should rise up before his eyes. Until then . . . he accepted everything as purely provisional. . . . He had given his allegiance to no theory, trembled for nothing, and longed for nothing; he readily rejected his own earlier convictions, for any new discovery might overturn all previous result.[16]

Certainly Brod studied Einstein closely during those Prague evenings. In Einstein he must have seen the "certain rigor and ruthlessness" he ascribed to Kepler.[17] When the novel was published in 1915, Einstein read it "with great interest," though with no certain memory of Brod himself. Others noticed the likeness: The German chemist Walther Nernst told Einstein: "This Kepler . . . that's you." [18]

In counterbalance to Einstein's "rigor and ruthlessness" was his sense of humor. Sometimes it came across as kindly, as when he wrote a child: "Do not worry about your difficulties in mathematics; I can assure you that mine are still greater."[19] But his humor was often darker, more biting. To the World Disarmament Conference of 1932, he offered this caustic message: "As it is, the hardly bought achievements of the machine age in the hands of our generation are as dangerous as a razor in the hands of a 3-year-old child."[20] Yet he remained cheerful. In 1915, he met the French pacifist and writer Romain Rolland in the safety of neutral Switzerland. The Kaiser was ruled by industrialists and mad generals, Einstein reported, and everyone in Germany looked forward to victory. Still, Einstein seemed "vivacious and serene" to Rolland:

> [H]e cannot help giving his most serious thoughts a jocular form. . . . Einstein is incredibly free in his judgments on Germany. . . . No German enjoys that freedom. Anyone else but he would have suffered from a sense of isolation in his thinking during this frightful year. Not he. He laughs. . . .[21]

In the face of isolation and hopelessness, Einstein mustered not only humor but fortitude, just as he would later through his long, lonely quest for a unified theory. Indeed, isolation seemed to suit him, on many levels. Most exiles regret having to leave their homeland. But Einstein considered his stateless state a blessing rather than a curse. When he became world famous, various nations tried to claim his allegiance. In 1919, he observed, wryly, in the London *Times:*

> By an application of the theory of relativity to the tastes of readers, today in Germany I am called a German man of science and in England I am represented as a Swiss Jew. If I come to be regarded as *a bête noire,* the descriptions will be reversed and I shall become a Swiss Jew for the Germans and a German man of science for the English![22]

"LIKE A DROP OF OIL ON WATER"

Einstein moved to Berlin just six months before the outbreak of World War I. Max Planck, whom Einstein revered, and Walther Nernst had come to Zurich in the summer of 1913 to discuss terms — or, as Einstein put it, they came looking for a "prize laying hen."[23] For a young man of thirty-four, the offer was generous: a triple appointment to the Prussian Academy of Science, the University of Berlin, and, as director, of the Kaiser Wilhelm Institute of Physics. The Kaiser Wilhelm Society was founded in 1911 with the goal of developing German science. Its research institutes in chemistry, physics, biology, and medicine were financed partly by the government, partly by wealthy businessmen. Inevitably, science became handmaiden to industrial and military aims, even before the Nazi takeover. Funding was plentiful and expectations ran high. By the time Einstein accepted his appointment, Berlin was arguably the center of the scientific world, though no match for the glamor and culture of Paris, Vienna, or London.[24] The Prussian Academy, by contrast, hearkens back to 1700, when Gottleib Liebnitz became its first president under Frederick, King of Prussia. Only two positions were fully salaried, and Einstein was offered one of them. Despite his childhood memories of rigid, militaristic German schools, he accepted. As his first public duty, he delivered an inaugural address on "Leibniz Day," July 2, 1914. One month later, war was declared.

Einstein was unusual in his revulsion toward the war. Even the respected and sober Max Planck took up the cause, encouraging students to enlist (the nearsighted Planck was exempt). As a Swiss citizen, Einstein was tacitly excused from rabble-rousing German nationalism. Nor did his reputation suffer particularly when he signed on to a petition by Georg Friedrich Nicolai, a renowned physician, protesting the war. Nicolai's reputation was ruined; Einstein was deemed eccentric. Having returned to Berlin as an outsider, he remained so politically. His professional and collegial relationships did not suffer. He was on friendly terms with the warmongers who surrounded him. He would not let the war and politics distract him.

In fact, he had no preparation for political activism. His adult life had been spent almost entirely on the sidelines in neutral Switzerland. His political education, such as it was, had consisted of listening to his socialist friends in heated conversation in Prague cafés. He was casually sympathetic. Nor had he been he an avowed pacifist before the war, though his contempt for German authoritarianism and exaltation of martial glory ran deep. He did not even like to play chess: too much naked competitiveness.[25]

To his surprise, his Berlin colleagues had suddenly transformed into superpatriots. Nor were they content to sit on the sidelines. On October 4, what became known as the "paper war" (*Krieg der Geister*) was launched with the infamous "Appeal to the Cultured World," drafted in October 1914 by the writer Ludwig Fulda.[26] The appeal was signed by ninety-three of Germany's most prominent scholars and artists; fifteen were scientists, including Planck, Fritz Haber, and Emil Fischer. The appeal condemned "enemies trying to befoul Germany's pure cause" in a struggle "forced" on it. It denied that Germany's invasion of Belgium was illegal and insisted that no violence had been visited on Belgian citizens. In a dramatic flourish worthy of Fulda's profession, it invoked the "legacy of Goethe, Beethoven, and Kant."[27]

As a Swiss national, Einstein was not asked to sign. But a growing sense of dismay led him to action. A few days later, he signed Nicolai's countermanifesto, "An Appeal to Europeans." Its premise was the international and transnational nature of science. It urged an end to the war without assigning blame. Nothing came of their international appeal. Only Einstein, Nicolai, an astronomer, Wilhelm Forster, and a Dr. Otto Beck could be persuaded to sign. Rather than publish immediately, Nicolai incorporated it into a study of war, which he published in Zurich in 1916.

In late 1914, Einstein joined the New Fatherland League (Bund Neues Vaterland), a liberal group trying to bridge war differences in Europe. They had connections in the German foreign office, but trouble arose when a statement of their peace views — signed among others by Einstein — was published in foreign newspapers in 1915. That same year, the League started distributing illegal writings by English pacifists — Bertrand Russell among them — and smuggling letters to pacifists in jail. The government finally shut down the League's offices, interrogated its officials, sent two female secretaries to jail, and forbade all other members to communicate with each other.[28]

In October 1915, the Berlin Goethe Society asked Einstein to write something for a "patriotic commemorative album." He submitted a three-page essay, "My Opinion of the War." The original essay spoke of the "sad circumstances of the present" and his hope that a European federation would rule out future wars. But it also denounced patriotism as a shrine to "bestial hatred and mass murder" and mockingly described a citizen's "affliction with a state to be a business affair, somewhat akin to one's relationship to life insurance."[29] The editors refused to print remarks so openly offensive to the Fatherland, and several paragraphs were excised. It says much about muzzled speech in Germany that Einstein's short essay was heavily censored by an organization bearing the name of Goethe.

Einstein was probably on the government's watch list of pacifists from the start of the war.[30] In 1915, he was elected to the council of the Central Organization for a Durable Peace and attended its closed-door session, the scholarly purpose of which was to make recommendations to diplomats about postwar matters.[31] The organization was hardly inflammatory or dangerous. Still, the German police investigated with pedantic thoroughness. Their reports included the name of the Berlin newspaper he subscribed to and a complaint that he had not always filled out the proper forms required to travel around Germany.[32]

Even though he was a Swiss citizen and a neutral, Einstein's antiwar stand might have led to his expulsion from Germany and from membership in the Prussian Academy. Dr. Nicolai, who drafted and signed the "Appeal to Europeans," was a German citizen, a professor at the University of Berlin, and on active duty as an office physician with the Army. But his outspoken pacifism played havoc with his life. He was "degraded to private and made a hospital orderly."[33] Einstein was far more prominent, yet he was never rebuked or even warned. The German authorities must have considered him annoying but relatively harmless.

Yet for a pacifist, Einstein's record was decidedly wayward. His criticism of Germany was most strident during the first year of the war. From mid-1915 to the Armistice, he said little publicly. His salary was partly paid by the industrialist Leopold Koppel; the Kaiser Wilhelm Institute, which he nominally directed, did military research. Einstein did not protest. (In England, Russell solved a similar problem of conscience by giving away his stock in a munitions firm to a financially strapped and decidedly not pacifist T. S. Eliot.) Einstein remained friendly with colleagues like Fritz Haber, who introduced poison gas into the war, and with the industrialist Walther Rathenau. Nor did Einstein refuse to serve in May 1916, when he was elected president of the German Physical Society, a loudly patriotic group.[34]

Surprisingly, Einstein even helped the German war effort by improving wing design for aircraft and gyrocompasses for U-boats. Of course, Germany in the First World War was not Nazi Germany; the Kaiser was a moral universe away from Hitler. Einstein, sooner than Russell, had also come to see the peace cause as a futile gesture against the prevailing "herd" mentality. The martial delusions of his fellow professors — bloodthirsty about the war but tender altruists at home — left him in mixed despair and amusement. As for the war itself, Einstein thought Germany and England both wrong, infected by an epidemic of lunacy.

Einstein was inconsistent, but neither obtuse nor hypocritical. He did not delude himself. In 1916, he confessed to a friend:

> Admittedly, things are fine for me here and I float on the very "top," but on my own and rather like a drop of oil on water, isolated by my attitude and concept of life.[35]

"On my own" and "isolated": What mattered to him was his work in physics. The rest — family, friends, love, worthy causes —

could be attended to only if there were time and energy. He was, after all, a young man in the grip of a discovery as great as any in scientific history. In November 1915, after more than eight years of torturous work, he was able to present to the Prussian Academy the definitive general theory of relativity.

ON THE PRECIPICE: BETWEEN THE WARS

By late 1918, with the war in its final throes, German sailors mutinied, refusing to fight and seizing ships from their officers. Revolution spread through port cities inland, eventually reaching Berlin. Soldiers and laborers rioted in the streets. A general strike was called. On November, the Social Democrat Philipp Scheidemann declared Germany a republic. Within hours, the Kaiser abdicated. Einstein, recovering from a debilitating ulcer, was elated by the prospect of a socialist and democratic government. Though he did not know it then, the most politically active part of his life had just begun.

Ironically, his entrance into the political fray followed on the heels of an extraordinary scientific event. On November 6, 1919, the Royal Astronomical Society in London reported that Einstein's prediction of light "bending" as it passed the sun had been proven. Arthur Stanley Eddington's photographs of an eclipse, taken in Sobral, Brazil, confirmed Einstein's calculations. The theory of general relativity was heralded around the world. On November 7, headlines touted the new sage: "Revolution In Science. New Theory of Universe. Newtonian Ideas Overthrown."[36] Einstein was thrust into the public spotlight and into a career forever dictated by notoriety.

The political value of his fame was obvious. The new German republic welcomed the left-liberal Einstein as a spokesman on its behalf. German scientists were typically conservative, yet the most illustrious scientist of all was ready to defend the shaky republic at home and abroad. He became the spokesman not only of the fledgling Weimar Republic, but of a passionate vision of internationalism.[37] In the first months after the Armistice, he was optimistic about both. In a 1919 letter to Max Born's wife, Hedwig, he wrote: "I believe in the growth potential of the League of Nations. . . . I don't believe that human beings as such can really change, but I am convinced that it is possible, and indeed necessary, to put an end to anarchy in international relations, even if it were to mean sacrificing the independence of various countries."[38]

One other cause beckoned him: Zionism. Ironically, given his abhorrence of nationalism, Einstein embraced Zionism passionately, though not without reservation. Zionism soon mattered more to him than anything outside of science. It was a surprising turn of mind. Until he moved to Berlin in 1914, he had little interest in being Jewish. The Jews he met in Prague in 1911 — Brod, Kafka, Hugo Bergmann — were committed Zionists, but they seem to have made little impression on him, at least in the short run: "I read the book [Brod's *The Redemption of Tycho Brahe*] with great interest," he wrote to Hedwig Born in 1916. "Incidentally, I believe that I met him [Brod] in Prague. I think he belongs to a small circle there of philosophical and Zionist enthusiasts, which was loosely grouped around the university philosophers, a medieval-like band of unworldly people. . . ."[39] Einstein's Jewish colleagues in Berlin were another matter: Many had converted to Christianity to become as German as possible, often with an eye to career advancement. The proudly independent Einstein called this servile "mimicry."

After the war, Zionism must have seemed a logical alternative to assimilation: "Judaism owes a great debt of gratitude to Zionism," he later said. "The Zionist movement has revived among Jews the sense of community."[40] He began to realize the utility of his fame as a scientist and as a Jew. Towards the end of the war, poor Jews fled Eastern Europe and poured into Berlin, crowding into an impoverished shantytown. To the dismay of assimilated, middle-class Jews, Einstein welcomed the refugees, but also urged them to look towards Palestine as the natural homeland of "free sons of the Jewish people." He further riled assimilated Berlin Jews in 1920, when he castigated the Central Association of German Citizens of the Jewish Faith for the underlying message of its name, so suggestive of a "servile attitude" and so resistant to ethnic identity. "Not until we have the courage to see ourselves as a nation, not until we respect ourselves, can we acquire the respect of others."[41]

In 1920, Chaim Weizmann, the driving force behind the Balfour Declaration of 1917 and its guarantee of a Jewish homeland in Palestine, became president of the World Zionist Organization. The following year, he set off on a trip to the United States to raise funds for a Hebrew University in Palestine. Einstein agreed to come along. "Naturally, I am needed not for my abilities but solely for my name, from whose publicity value a substantial effect is expected among the rich tribal companions in Dollaria."[42] For all his arch deprecation, he was proud to lend his name to the cause. Howls of outrage were hurled at him for taking such a public stand in favor of Zionism, and especially for setting foot on the soil of Germany's enemies (he stopped in England on the way home from the United States). Before setting out, Einstein heard from his colleague Fritz Haber: "If at this moment you demonstratively fraternize with the British and their friends, people in this country will see this as evidence of the disloyalty of the Jews."[43] Still, Einstein made the trip. He was greeted in the United States with such enthusiasm as would gratify a movie star. This was Einstein's first visit to the country, an exhausting, sometimes preposterous experience. The endless interviewing, hand-shaking, and touring, the crowds gaping to catch sight of him, reporters asking inane or sensationalizing questions, and he answering with "cheap jokes" taken seriously — Einstein said he felt like a "prize ox" being exhibited. In a spare moment, he made his first trip to Princeton as well, receiving an honorary degree from Princeton University and giving four lectures there on relativity (published as *The Meaning of Relativity*).

Meanwhile, he dutifully worked for the Weimar Republic. He served on a committee to evaluate German war atrocities; he joined the League of Nations' Commission on Intellectual Cooperation. He traveled widely, representing (despite his Swiss citizenship) German science and the democratic German government, in sometimes controversial efforts to overcome the animosities of the war. In the spring of 1922, the Collège de France asked him to lecture in Paris. After some hesitation, he accepted. He was the first German scientist to be invited, though many French scholars disapproved, and Einstein's German colleagues were equally unhappy.[44] Increasingly certain of his commitment to internationalism, Einstein turned his fame into a pan-national passport.

Then politics turned dangerous. His friend Walther Rathenau, a Jewish industrialist with philosophic inclinations who became foreign minister in early 1922, was assassinated in June. Political murders in postwar Germany were common. One estimate reported that in 1922, left-wing death squads were responsible for twenty-two such killings; right-wing death squads were responsible for more than three hundred.[45] As a Jew, a liberal, and an internationalist, Rathenau was anathema to the conservatives, diehard militarists, anti-Semites, and nascent Nazis making up the political right.

Einstein knew that he was in danger. His defiant Jewishness, his antiwar activities, his "Red" sympathies during the German revolution in 1918, and especially his world fame — everything pointed to his becoming a target of the right. He canceled his lectures and political work, and even considered resigning from the Prussian Academy for the first time since 1920, when an antirelativity rally left him wondering whether to leave Germany altogether.[46]

It was time for him to lower his too-familiar profile. He and Elsa sailed to Japan, where he had been invited to give lectures. Oddly enough, Bertrand Russell had made this trip possible, perhaps unwittingly. During his own trip to Japan in 1921, Russell had been asked by a publisher to name the most "significant" people alive. His answer: Lenin and Einstein. The Japanese publisher decided on Einstein (Lenin having his hands full with the Russian Revolution). Not only did the lecture series allow Einstein to absent himself from the dangers of Berlin; it also contributed handily — a whopping £2,000 in British currency — to his constantly depleted bank account. He spent six weeks in Japan, lecturing to crowded rooms while his words were painstakingly translated into Japanese.

As he sailed back, en route to a planned stop in Palestine, he learned that he had won the Nobel Prize in Physics. Despite the news, Einstein continued his trip to Jerusalem. He was the honored guest of the British High Commissioner of Palestine, Sir Herbert Samuel, a fellow Jew.[47]

The Nobel Prize had long been a certainty, with only one question in the air: Why had it taken so long? The quandary faced by the Nobel committee had much to do with Einstein's eminent qualifications for the award. He had been repeatedly nominated for his major works: the photoelectric effect, Brownian motion, and relativity. The latter, perhaps the most logical choice for the award, evidently ran up against a seeming technicality: The Nobel Prize is awarded for a discovery, not a theory. More to the point, despite definitive proof by Sir Arthur Stanley Eddington, relativity stirred controversy. In Germany, promulgators of the nascent "German Physics" (including the rabidly anti-Semitic Paul Weyland and Ernst Gehrcke) denounced the theory. In Sweden, the committee members had difficulty understanding it. In the end, it was decided to give the award to Einstein not for relativity, but for his discovery of the photoelectric effect. Thus, the paper he called "revolutionary," the first published paper of his "miracle year," was the feat for which Einstein won his Nobel.

By the terms of their divorce agreement, the prize money went to Mileva. (She continued to care for their two sons, and especially for Eduard, whose mental health was increasingly fragile.)

Ironically, the awarding of the prize led to a small international quarrel. Germany wanted to claim this latest laureate as its own, insisting that members of the Prussian Academy were automatically German citizens — thus, Einstein was deemed to be German, despite his Swiss citizenship. Einstein objected vehemently. Since protocol required that the winner's national representatives take part in festivities, the quarrel was more than academic. Who would deliver the medal to Einstein? Finessing the problem, Nobel Foundation officials dispatched a Swedish minister to Einstein's apartment, where he handed over the honorific medal and scroll.

Though adamant about his Swiss citizenship, Einstein had always claimed to be a man without a homeland — a member not of a nation, but of the international community. But during the 1920s, he began to identify strongly with Zionism and Jewish causes. It was the closest he ever came to nationalism. Though passionate in his feelings, Einstein never hesitated to criticize the nascent Jewish state. Hebrew University in Jerusalem enjoyed his early support, but by 1928, despairing of its quality, he resigned from its board.[48] He repeatedly encouraged "peaceful cooperation" with Arabs, blaming British policy for the deepening and dangerous animosity. Yet his love for Israel and his identification with Jewish causes never wavered. His feelings were enthusiastically returned: From his sudden fame in 1919 to his death, Einstein was seen by most of the world's Jews as their greatest living figure. When Chaim Weizmann died in 1952, David Ben-Gurion, then prime minister of Israel, decided to offer the presidency to Einstein:

> There is only one man whom we should ask to become President of the State of Israel. He is the greatest Jew on earth. Maybe the greatest human being on earth.[49]

Quite sensibly, Einstein refused. Still, he was deeply moved by the offer from "our state Israel":

> All my life I have dealt with objective matters, hence I lack both the natural aptitude and the experience to deal properly with people and to exercise official functions. . . . I am the more distressed over these circumstances because my relationship to the Jewish people had become my strongest human bond, ever since I became aware of our precarious situation among the nations of the world.[50]

It now requires an effort of historical imagination to recall how much dignity Einstein lent to Jews. Supreme in science, manifestly decent, he was a living refutation of racial and religious slurs. That such a great mind saw himself as a Jew like all the rest; that he said so at a time when millions of poor Jews still lived in Eastern Europe or congregated as immigrants in the slums of New York or Berlin, often resented by their more fortunate brethren; that he spoke out tirelessly to defend them or attack their enemies — these efforts made him not only admired but beloved by his fellow Jews. When Hitler took power in 1933, Einstein's immediate denunciation of the Nazi regime carried powerful weight around the world. His early support of Zionism was of incalculable value, and Weizmann knew it, though he was often annoyed by Einstein's naïveté or obstinacy. In 1918, when Einstein became a Zionist, that cause was largely ignored or unpopular among the mass of Jews; a few thousand emigrated to Palestine, many millions to America and other Western countries. He was the "Jewish saint," Einstein said ironically, but he never shirked the responsibilities involved. Russell was born into the ruling class of the most powerful empire on earth — Einstein the Zionist was an early patriot of a nation that did not even exist until 1948.

For Einstein, the years between the wars saw the diminution of his scientific genius. They saw, as well, a new focus on the atom. Relativity had resolved macrophysics — the realm of gravity, time, and space. What remained was the invisible realm of what makes matter.

In some ways, Einstein remained at the center of physics purely by dint of his reputation. Throughout the late 1920s and early 1930s, he reestablished his status as an outsider for his stubborn rejection of quantum mechanics. A younger generation, located in the "quantum triangle" of Munich, Göttingen, and Copenhagen, succeeded in changing our worldview, as he had in relativity.[51] It was, in many ways, a more startling revolution than Einstein's. Quantum mechanics gave us the structure of the atom, but it robbed us of the certainty of causation. Einstein was never reconciled to a physics in which God appeared to "throw dice." When, in 1932, he finally accepted the usefulness and (for him, limited) success of quantum mechanics, he turned irrevocably away from mainstream physics. His search for a "theory of everything" based on relativity, a unified theory that would subsume quantum theory, was relentless and, ultimately, unsuccessful.

Fame assured Einstein of the means to carry on his work. As Germany descended into the hell of Nazism, he found himself afloat in requests for lectures and job offers. He lectured at the new California Institute of Technology in 1930 and again in 1932. He was invited to deliver the Rhodes lecture at Oxford in 1931. In 1932, while at Caltech, he met Abraham Flexner, an academic reformer and inveterate organizer, who was recruiting faculty for his newly endowed Institute for Advanced Study in Princeton. Flexner, determined to entice the greatest name in science, offered Einstein a six-month contract scheduled around his Berlin duties. But Einstein was as yet unwilling to abandon his home in Berlin.

Almost immediately after Hitler took power, Einstein was targeted. That the most famous scientist in the world was a Jew — not to mention a pacifist and internationalist — outraged the Nazis. In the years leading up to Hitler's election, Einstein had attended rallies, signed manifestos, and lent his name to appeals in the anti-Fascist cause. Once in power, the Nazis wasted no time in retaliating. Among the photographs in a book listing "traitors to

Germany," Einstein's picture was captioned: "Not Yet Hanged."[52] Soon to come were book burnings, mass dismissals of Jews from academia, and concentration camps.

Einstein did not hesitate. Only three months after the Nazis took over, he cut all ties with Germany, resigning from the Prussian Academy of Sciences and turning in his passport.[53] The Nazi chieftains and German newspapers spat at him, calling him a turncoat. As Einstein wrote to Max Born, "I've been promoted to an 'evil monster.'"[54] The Prussian Academy — with a very few honorable exceptions — rejected Einstein's resignation. They wanted the satisfaction of expelling him, and did so a few months later. Flexner's offer of a half-year stay was extended indefinitely. As it turned out, Einstein remained at the Institute for the rest of his life. He never set foot in Europe again.

For almost its first decade, the Institute for Advanced Study was an institution on paper only, with no buildings of its own. Offices were rented for its members from Princeton University. Its purpose was to free its scholars from teaching. Lavish salaries also made it attractive: Einstein asked for a salary of $3,000, but was given $15,000 per year — this during the early Depression, when most American professors earned about $2,000. That salary was possible because the Institute was handsomely funded by Louis Bamberger and his sister, Caroline Bamberger Fuld, who had sold their department store in 1929 a few scant weeks before Wall Street crashed.

Einstein was the second professor named; the first was the American mathematician Oswald Veblen. Until 1935, six mathematicians comprised the entire fulltime Institute — but they were a choice group: Besides Einstein and Veblen, they included the renowned mathematician Hermann Weyl, self-exiled from Göttingen, and the young John von Neumann. As a bonus, one of the visiting scholars in 1933 was the twenty-seven-year old Gödel.

If not for his fame, Einstein would very probably not have

been hired. Mathematicians were the first appointees because there was consensus about the best. Einstein was never a mathematician in the sense of a Gödel or a von Neumann; he was a physicist who used whatever abstruse mathematics he needed, but beyond that his interest and expertise dropped off. Still, general relativity had spurred new mathematical research, and he was in any case too valuable a catch to lose.

On his way to Princeton in October 1933, Einstein sailed to New York, to be met there by the mayor, a band, speeches, and the usual journalistic hoopla. But Institute officials, worried about conservative protests against Einstein as a Bolshevik, hurried him off to Princeton.[55]

So, at age fifty-four, Einstein settled in a most unlikely place: a small American college town nestled amid genteel wealth. Nothing could have been more different from Berlin, where art, decadence, and scientific eminence jostled with Nazis and Communists bloodying the streets, and where Hitler now reigned. The leafy Princeton streets quietly shaded an inbred little community, affluent and above all decorous. Princeton University embodied these qualities in its neo-Gothic architecture reminiscent of Oxford. From about 1920, the university's mathematics department had suddenly blossomed into one of the greatest centers of mathematics in the world, doing new research in every direction. But Fine Hall, which housed it, looked backward, at least architecturally, to old Europe and the Gilded Age of nineteenth-century America: its long corridors punctuated by stained-glass windows, its offices carpeted and lavishly furnished.[56] Oswald Veblen, said to have planned the building, was the nephew of Thorstein Veblen, the famous American social thinker who satirized "conspicuous consumption" — spending lavishly to excite the envy of others — which might describe the Fine Hall of Mathematics. The corrosively ironic Thorstein Veblen was one of Einstein's favorite authors, along with Russell.

EINSTEIN AND RUSSELL: PARALLEL LIVES

Einstein's last letter, written within days of his death in 1955, was addressed to his friend, Bertrand Russell:

> Thank you for your letter of April 5. I am gladly willing to sign your excellent statement. I also agree with your choice of the prospective signers.
> With kind regards, A. Einstein[57]

This short note was a fitting last word to a lifelong friend. It added Einstein's name to what became known as the Einstein-Russell Manifesto. The manifesto was signed by nine other scientists, among them Max Born, Linus Pauling, and Frédéric Joliot-Curie. Conceived by Russell, it called upon the American Congress, and the public, to repudiate war in the face of nuclear weapons. It met with a surprising degree of support, despite the Cold War. Certainly Einstein's signature, offered on his deathbed, lent prestige and credibility. Only days after its publication, an industrialist named Cyrus Eaton offered to fund the conference proposed by the manifesto, providing that it take place in Pugwash, Eaton's birthplace in Nova Scotia. Thus were the Pugwash Conferences and other anti-nuclear movements born.

Among public intellectuals with international stature, few were more visible in their antipathy for nuclear arms than Einstein and Russell. The two had been comrades in pacifism since World War I, when Russell in England and Einstein in Germany were the most prominent figures on either side to speak out against the slaughter. Though the two men saw little of each other throughout their lives, they lived and thought along similar lines — outspoken defenders of peace, social justice, and intellectual freedom.

As it had done for so many, World War I changed both their lives. Russell plunged into political action, gathering signatures from Cambridge dons, writing an antiwar letter to *The Nation,* and joining the Union of Democratic Control and the No-Conscription Fellowship.[58] In Germany, Einstein, who had just moved to Berlin, was appalled by the rabid nationalism of leading German intellectuals. He signed the antinationalist "Appeal to Europeans" and, for the first time, joined a political association, the Bund Neues Vaterland (New Fatherland League).

If their scientific energies had relatively short half-lives, their political energies, once unleashed by the two World Wars and their interregnum, knew no boundaries.

The two men first met sometime in the early 1930s, possibly when Russell stopped in Princeton during an American tour. Their first known correspondence seems to have been in 1931, when Einstein wrote, expressing his "highest admiration" for the mathematician turned philosopher.[59] (Russell returned the compliment, calling Einstein "the leading intellect of the age."[60]) Had Plutarch lived today, he might have chosen Einstein and Russell as subjects of a modern *Parallel Lives*. By education and outlook, they were Victorians; by dint of genius, they helped define modernity. They shared the high optimism of purpose and confidence bequeathed by the European empire and the skepticism bequeathed by its inexorable demise. Both worked energetically throughout their lives. They carried on the Victorian practice of reading aloud to family. Einstein, for one, read to his invalid sister each evening from Herodotus or Xenophon — or Bertrand Russell. After his first marriage in 1894, the twenty-two-year-old Russell and his bride happily entertained each other in the evenings by reading aloud from Shakespeare, Gibbon, Plutarch, and Shelley's *Epipsychidion*.

The two men also shared an intellectual heritage. Each had discovered Kant early in life (Einstein in his early teens; Russell at Cambridge).[61] John Stuart Mill and David Hume were foundational, especially Mill for Russell. In later years, Einstein remembered Hume's works as having had "considerable effect on my development."[62] (Einstein's admiration of Russell's wit, logical precision, and skepticism may have its roots in the resemblance to

Hume.) Although canonized, along with Nietzsche, Hegel, and Freud, as arch-shapers of the modern spirit, Einstein and Russell both preferred the cool rationalists of the seventeenth and eighteenth century. Russell paid scant attention to such twentieth-century fevers as nihilism, existentialism, or the unconscious psyche. Einstein bothered with them even less and had no more to do with moral or historical relativism than the accident of the word "relativity." *His* theory sought to rid physics of subjective relativity; indeed, the term "relativity theory" was coined by Planck, who nevertheless marveled over the "absolute, the universally valid, the invariant" uncovered by the theory. "Invariance theory" was proposed at one point as an alternative name, but "relativity" stuck.

Above all, they were committed leftists, whose politics were tempered by inconsistencies. Einstein, a lifelong enemy of nationalism, supported Zionism and the creation of Israel — "I am against nationalism but for the Jewish cause" seemed to be a compromise he could live with. He championed freedom and equality but also declared that "the creative sentient individual . . . alone creates the noble and the sublime, while the herd remains dull in thought and feeling."[63] Russell was a political radical who remained an aristocratic with an aristocrat's sense of privilege. Ever at odds with the spirit of his times, he found himself drawn to and ultimately abandoned by young turks (Wittgenstein and D. H. Lawrence, especially) because he could not or would not subscribe to their views. The First World War moved Einstein and Russell into collision with politics and changed their lives. By the 1920s, they were activists and global celebrities, Einstein the Sage of the Universe and Russell the nonconformist founder of analytical philosophy.

And yet, few men were so different, above all, in their heritage. Einstein was born into a middle-class family, but as a Jew and, for most of his life, a man without a country, he was very much the outsider. The aristocratic Russell was by birth an insider. Their given names reflect social class. Einstein's parents straddled the awkward, tentative path of assimilation, giving Albert a nonbiblical name, but not the more assimilated "Albrecht." "Bertrand" had the opposite purpose — to emphasize distinction. The Russell family gained prominence when Henry VIII replaced the troublesome Catholic aristocracy with loyal Protestants. He ennobled John Russell and gave him splendid estates and abbeys. Bertrand's grandfather was an earl, a title Bertrand inherited in 1931.

In character and demeanor, too, the differences were stark. One has only to think of the famous photograph of the elderly Einstein sticking his tongue out at the camera — childlike impudence mixed with self-mockery, the celebrity scoffing at anyone foolish enough to take celebrity seriously. Earthy and droll, Einstein also tended to be gentle with friends. More often than not, he was the target of his own sarcasm: "So now I too am an official member of the guild of whores," he said upon taking the Patent Office position.[64]

Such antics were never Russell's style. If Einstein was the benign clown, Russell was the witty satyr. T. S. Eliot, who had been Russell's student at Harvard in 1914, caricatured Russell in "Mr. Apollinax" as an interloper at a tedious faculty party, "laugh[ing] like an irresponsible foetus" and "grinning over a screen / With seaweed in his hair." He did not fancy the Harvard dullards, and he let them know it. His wit was often used to flay. Of the Bloomsbury set, he snarled:

> They put up with me because they know I can make anyone look ridiculous — if I had less brains and less satire, they would be all down on me — as it is they whisper against me in corners, and flatter me to my face. They are a rotten crew.[65]

One hardly knows whether to be awed more by the ruthless candor or by the ruthless turn of language. The power to sting never deserted Russell. Throughout his life, he discharged his quick brains and combative wit against the enemy, whoever it might be. When the target was public injustice or folly, Russell could be a refreshing gadfly — witness his early courageous attack on Soviet oppression in 1920, which earned the wrath of many intellectuals. Of his travels in Russia, he wrote, "I felt that everything I valued in human life was being destroyed in the interests of a glib and narrow philosophy, and that in the process untold misery was being inflicted upon many millions of people."[66] In private company, his acuity could make him chilling and feared: "He has the tongue of a witty, acidulous and far from benignant adder," wrote Leonard Woolf in 1968, noting not only Russell's tendency to flay, ·but his prejudices as well — his "dislike and hatred of Americans, Jews, and even his personal friends."[67] He seemed to fascinate and repel in equal measure. D. H. Lawrence lampooned him in *Women in Love* (1920) as a "learned dry baronet" with frozen feelings and a "harsh horse-laugh." Aldous Huxley cast Russell as Mr. Scogan in *Crome Yellow* (1921), harping on his "rather fiendish laugh":

> Mr. Scogan was like one of those extinct bird-lizards of the Tertiary. His nose was beaked, his dark eye had the shining quickness of a robin's. But . . . his hands were the hands of a crocodile. His movement was marked by the lizard's disconcertingly abrupt clockwork speed . . . an extinct saurian.[68]

Russell often struck observers this way: beak-nosed, small and bony, quick as a bird in response — and in his ability to peck. Einstein's sad eyes and wild hair seemed those of a simple saintly genius, a poetic dreamer. These familiar images distort both men. The "icy" Russell was as often stirred and generous; Einstein was neither simple nor saintly.

Admiration and praise of another expresses one's own inner self. Einstein's heroes and heroines were many. Foremost was Hendrik Lorentz, the great Dutch physicist, whose "quite unusual lack of human frailties never had a depressing effect on others."[69] Max Planck was another father figure with whom Einstein remained friendly, despite their political disagreements. Of Marie

Curie, Einstein recounted "twenty years of unclouded and sublime friendship":

> Her strength, her purity of will, her austerity toward herself, her objectivity, her incorruptible judgment. . . . [O]nce she had recognized a certain way as the right one, she pursued it without compromise and with extreme tenacity.[70]

Einstein had the same tough qualities, though unlike Curie, he was constitutionally cheerful. Madame Curie, he once lamented, could not shake off sadness: "She had the soul of a herring."[71] Not Einstein.

Russell often saw himself reflected in those he admired, as if their genius might shore up his self-confidence. He felt a special kinship with the novelist Joseph Conrad. "In the out-works of our lives, we were almost strangers, but we shared a certain outlook on human life and human destiny, which, from the very first, made a bond of extreme strength." Conrad's pessimistic "philosophy of life" led him to "a profound belief in the importance of discipline . . . by subduing wayward impulse to a dominant purpose."[72] Russell, too, gazed into the abyss, and his life was a ceaseless struggle to subdue that "passionate madness" to which he, like Conrad's characters, tended. Russell never gave way to madness, but he was tormented by it, and by the passions to which he confessed in the first sentences of his *Autobiography*:

> Three passions, simple but overwhelmingly strong, have governed my life: the longing for love, the search for knowledge, and unbearable pity for mankind. These passions, like great winds, have blown me hither and thither, in a wayward course, over a deep ocean of anguish, reaching to the very verge of despair.[73]

Such buffeting by warring passions is nowhere visible in Einstein. In his soul there was no abyss. He remained grandly, classically optimistic: "The eternal mystery of the world is its comprehensibility."[74] In 1949, he composed some "Autobiographical Notes" for a volume in his honor. In sharp contrast to Russell's "ocean of anguish" (although Russell, too, indulged in gallows humor), Einstein's reminiscences begin with a humorous and jaunty bit of self-deprecation: "Here I sit in order to write, at the age of sixty-seven, something like my own obituary."

RUSSELL: ARISTOCRAT IN TURMOIL

Bertrand Russell never seemed to lack social confidence. Rupert Crawshay-Williams, a young friend, describes a trip with Russell to Wales to check on a house that was being renovated, in Russell's view, too slowly. As they drove, Russell calmly discoursed on philosophy with Williams and his wife, Elizabeth. At the house, they found both builder and architect. Without more than a perfunctory greeting, Russell began to rant:

> "What do you mean by this intolerable and quite inexcusable delay?" he roared. The builder and the architect were so taken aback by this eruption that they were speechless for the first few minutes. They went pale with astonishment and their lips trembled. . . .[75]

Russell railed on and on. The workers were left "gasping and floundering." Then he strode out to the car and calmly resumed the conversation as if nothing had happened. His companions were astonished. Might the workers now think badly of Russell, ventured Crawshay-Williams, having heard so much thunderous criticism? "No, I wasn't worried about that. Why should I be?" Russell the aristocrat knew his place in the social strata and did not hesitate to trade on it.

But the inner Russell was not quite so confident. Much of his insecurity was fostered by a difficult childhood. Born into an aristocratic, unconventional family of stalwart Whigs, Russell inherited his liberal politics and, for better or worse, his Victorian sensibilities. His mother and father died before he was three, after which he and his older brother Frank were sent to live with their paternal grandparents at Pembroke Lodge. Lord John, the boys' grandfather, once a monument of nineteenth-century English history (having launched the great Reform Bill of 1832 and served two terms as prime minister), died in 1878, when Russell was just six. Lord John's widow, Lady Russell, though twenty years younger than her husband, was formidable and utterly unsuited to parenting.

Lady Russell was widely read, perfectly fluent in German, French, and Italian, and fearlessly unconventional. She was also immensely puritanical. For the good of her soul, she eschewed wine, meat, even the luxury of a "comfortable chair" for most of the day.[76] Her self-abnegation extended beyond herself to her grandsons. The young Bertrand was forbidden even modest pleasures: No fruit, only the rare piece of candy, cold baths, and a rigid schedule were the order of each day.[77] At Pembroke Lodge, Lady Russell's great house, a succession of eccentric relatives and eminent personages came and went.

So, too, at the more vibrant household of the sharp-tongued Lady Stanley, his maternal grandmother. Russell remembered the "hawk's eye" of Prime Minister Gladstone and, at a time when the Irish nationalist Charles Stewart Parnell was under suspicion of murder, a succession of Irish MPs.[78] Bertrand's godfather was John Stuart Mill, long a family friend. All around the boy, the talk was of high politics and Whiggish history. It was Lady Russell's great hope that Bertrand would enter politics, as his father and grandfather had before him.

The boy had no real friends and few companions his own age. Frank escaped the dreary house early by waging a campaign of misbehavior until he was finally sent away to Winchester. Not Bertrand, who was tutored at home. He sometimes played games with the servants when their chores allowed. A solitary, brilliant child, he grew prematurely intellectual, analyzing others and himself. At eleven, he fell in love with Euclid. In his teens, introspective, precocious, and shy, he naturally kept a diary, but wrote his thoughts in secret Greek characters.

Trinity College at Cambridge was his salvation. When he entered at eighteen, he was astonished to find other people like himself. He was not as odd as he had feared. A. N. Whitehead, Crompton Llewelyn Davies, Ellis McTaggart, Robert and Charles Trevelyan, Roger Fry, Lowes Dickinson — these young men were friends and colleagues with whom he could share the "whole world of mental adventure."[79] The shy, priggish, slight adolescent was, by the time he left Cambridge, self-confident and, if not worldly, at least part of a world.

Russell's "first" in mathematics from Trinity was followed by a "first" in philosophy (Moral Sciences) in 1894. Immediately upon graduating, while working at the British embassy in Paris, Russell traveled to Germany, where he attended party meetings of the Social Democrats, who had just been outlawed. The trajectory recapitulates his life — one devoted first to mathematics, then to philosophy, and finally to left-leaning political action.

In 1893, Russell began a romantic correspondence with Alys Pearsall Smith, the daughter of American Quakers living in Surrey. Alys was a serious, pious social reformer four years older than Russell. He fell in love immediately, but prudently waited until he was twenty-one, at which time he came into his inheritance and felt sufficiently independent to brave his family's inevitable disapproval. His formidable grandmother seems to have rejoiced when, momentarily, the marriage was delayed by fears that insanity in both families might be passed on to their children.[80] The crisis was resolved when they assured each other (falsely, as it turned out, on Russell's side) that children were not necessary. The specter of birth control, which the Russell family doctor deemed "injurious to health," was another impediment. At last, a second opinion laid to rest Russell's qualms. They married in 1894. Their marriage lasted until 1911, when Russell fell in love with Lady Ottoline Morrell. He and Alys were finally divorced in 1921.

His academic life began when he was appointed a lecturer at the London School of Economics in 1896. By then, he had published *German Social Democracy*, his first book. Thereafter, of course, he evolved into the familiar Russell of the modern landscape: The archrationalist brushing away logical cobwebs, metaphysical confusion, and the shackles of religion, and the champion of social progress. He was the new Voltaire. Yet within, he always felt different and apart. His puritan impulses were at constant war with his desire for sexual liberation, and his yearning for certainty never seemed to square with the rationalism to which he aspired. Above all, his long experience with solitude in childhood haunted him, as did fears of insanity. Russell championed mathematical logic and analytic philosophy. He was, said Ottoline, "so quick and clear-sighted, and supremely intellectual — cutting false and real asunder. Somebody called him the Day of Judgment."[81] Yet he tangled incessantly with the vagaries of passion: "[M]y nature is hopelessly complicated; a mass of contradictory impulses."[82] These impulses, so beyond his control, led to a lifelong passion for self-analysis.

From boyhood on, Russell recorded his inner self in letters, memoirs, journals, essays, and books. No other important philosopher has recorded his own life in such detail. Much of this detail is self-deprecating, all of it extremely revelatory, nowhere more so than in his *Autobiography*, the final volume of which was published just before his death. Writing one's own life, observes Michael Foot in his introduction to the *Autobiography*, "is the most risky and arduous of all the writer's arts."[83] It is courageous in great part because whatever it reveals can be taken at face value or deemed disingenuous — either way, in the effort to set history "right," the autobiographer inevitably provides more ammunition.

Russell's *Autobiography* alternates narration with contemporary letters to and from him. He scrutinizes his sexual maturation and its vicissitudes: his awkward masturbation as a youth; the enduring "comic" difficulties of coitus with his first wife, Alys; learning from his lover Lady Ottoline Morrell of his pyorrhea, finally cured by an American dentist. He records humiliation and anguish at the hands of his onetime pupil Wittgenstein and his onetime acolyte, the writer D. H. Lawrence.

His relationship with Ottoline, the first truly uncontrollable love affair of his life, set off equally uncontrollable behaviors. Angry with her once, he walked twenty miles in the rain, "in a fit of madness."[84] Sexual passion left him haunted by the fear of insanity: "It doesn't do for me to relax too much — the forces inside are too wild — some of them must be kept chained up."[85]

Insanity must have seemed a threat. He was plagued by nightmares "in which I dream that I am being murdered, usually by a lunatic." One night, he awoke from a nightmare with his hands around Alys's throat.[86] He was no stranger to rage: At eighteen, he nearly choked his best friend to death; he attempted to smother his third wife, Patricia, with a pillow; he felt, in his own words, "murderous impulses" toward Paul Gillard, an acquaintance whom he called a "drunken homosexual spy." He wasn't afraid of "peccadilloes," he said, but he was terrified of "big violent crimes — murder and suicide and such things."[87]

Self-dramatizing was a mode of operation for Russell, especially with women. To them he would reveal these impulses: "[M]ost people would despise my inner turmoil." Yet his tone was clinical. "Only intellect keeps me sane: perhaps this makes me overvalue intellect as against feeling."[88] Ray Monk argues that Russell sought to control his frightening impulses by dint of cold reason.[89] Although Russell records no "murderous" episodes or feelings after his fifties, his wish to douse feelings with intellect never deserted him, to the extent that, as Crawshay-Williams argues, his public persona of "materialist temperament and unfeeling intellect" obscured his "sympathetic emotions."[90]

Russell also had uncontrollable impulses of a more benign sort. He called them "conversions,"[91] since they struck him with the force of a mystical experience, bringing him an understanding that his intellect could not achieve.

His relationship with Ludwig Wittgenstein was one such "conversion." Wittgenstein was a wealthy but unknown young Austrian who came to study logic with Russell in 1911. Driven by his search for truth, caring nothing for feelings or consequences as he pursued his own vision, the twenty-two-year-old Wittgenstein began to consume the thirty-nine-year-old Russell. He often appeared in Russell's rooms at midnight, pacing the floor as if caged, silently. He harangued Russell hour after hour with philosophic discourse. He raged at those who claimed not to understand him. He threatened suicide. He was impervious to any ideas except his own.

Russell, always generous when he saw intellectual fire and interest, put up with and even encouraged his strangely irrational and brilliant disciple. In only a few months, Russell came to see himself as the disciple. "[Wittgenstein] has more passion about philosophy than I have; his avalanches make mine seem mere snowballs." The conversion was complete. "Wittgenstein has been a great event in my life . . . I think he has genius. . . . I love him & feel he will solve the problems I am too old to solve."[92] Yet two years later, Wittgenstein pronounced the manuscript of Russell's new book, *Theory of Knowledge,* "all wrong," demolishing the thesis in a series of devastating exchanges. Now it was Russell's turn to think of suicide.[93]

This episode is often viewed by Wittgenstein's adherents as a clash between youth and aging genius. In many ways, Russell agreed. By the end of their relationship, Wittgenstein had convinced the older man that he "could not ever hope again to do fundamental work in philosophy."[94] Still, what overwhelmed Russell more than Wittgenstein's intellectual agility was his imperially romantic personality. The self-absorbed young Austrian either dominated those around him or banished them from his sight — a "tyrant," Russell called him.[95] Yet Russell felt that Wittgenstein had "a kind of purity. . . . [H]is personal force was extraordinary." By force of personality, the youthful Wittgenstein swept Russell away: "He was perhaps the most perfect example I have ever known of genius as traditionally conceived, passionate, profound, intense, and dominating."[96] Russell had other philosophic opponents during his life: Henri Bergson, William James, G. E. Moore, F. H. Bradley, John Dewey, the later Whitehead, and Henri Poincaré. But he never allowed their criticism to affect him as did Wittgenstein's.

The same intensity characterized his brief friendship with the novelist D. H. Lawrence. They met in 1915, when Lawrence was twenty-six and Russell forty-three. Lawrence was convinced that a wholly new social order was necessary to prevent more wars. Russell, tired of "Sunday-schooly" pacifists, welcomed Lawrence's iconoclasm. Having taken up the cause of pacifism with a vengeance, Russell hoped that the fiery Lawrence "could give me a vivifying dose of unreason."[97] He was looking for a prophet, and he found one in Lawrence. After their first talk, Russell was enthralled. He described his new friend to Ottoline Morrell:

> "[Lawrence] is amazing; he sees through and through one."
> "Yes. But do you think he really sees correctly?" I asked.
> "Absolutely. He is infallible," was Bertie's reply. "He is like Ezekiel or some other Old Testament prophet . . . he sees everything and is always right."[98]

Russell took Lawrence to meet his cabal of Cambridge friends: the economist Maynard Keynes, the classicist Lowes Dickinson, and G. E. Moore. A few years earlier, Wittgenstein had denounced Russell's Cambridge circle as utter fools and derided Russell for being polite to them. Lawrence, likewise, "hated them all with passionate hatred and said they are 'dead, dead, dead.'" At first, Russell did not disagree, thinking that such an imaginative genius had "an insight into human nature deeper than mine." But he was increasingly repelled by Lawrence's mystical philosophy of "blood consciousness" — which, as opposed to intellect, was that aspect of "consciousness . . . belonging to darkness."[99]

The two men fell out when Russell argued the unexceptionable view that people were capable of kindly feelings towards one another. Lawrence responded by letter, furiously accusing Russell of hypocrisy and cowardice. Like everyone else, Lawrence argued, Russell only wanted to satisfy his "lust to jab and strike." He scornfully advised Russell to give up peace work and stick to sterile mathematics. As he had been after the quarrel with Wittgenstein, Russell was devastated.

> I was inclined to believe he had some insight denied to me, and when he said that my pacifism was rooted in blood-lust I supposed he must be right. For twenty-four hours I thought I was not fit to live and contemplated suicide.[100]

Then he came to his senses. "One must be an outlaw these days, not a teacher or preacher," Lawrence had earlier written him. But Russell quite rightly reflected that "I was becoming more of an outlaw than he ever was."[101] His infatuation with yet another dominating romantic genius ended; their friendship had lasted a year.

Russell did not stick to "sterile mathematics," and his contributions to original philosophical thinking virtually ceased by the early twenties. He returned, briefly, to so-called "technical" work only once, in 1924, when he revised the *Principia Mathematica* for a second edition. But the bulk of his postwar work lay in politics and popular writings. To a lay audience hungry for knowledge, he wrote highly accessible books on relativity, the atom, and twentieth-century philosophy.

Russell's wide-ranging interests and radical political activities were tame in comparison with the tumult of his emotional life. He was married four times, first to Alys Pearsall Smith, whom he married in 1894, then to Dora Black, who was pregnant with Russell's first child when they married in 1921. Russell and Black divorced in 1935. They had two children together — John and Katherine. After enduring their parents' messy divorce, the two children found themselves with a difficult stepmother, Marjorie (Peter) Spence, whom Russell married in 1936. Conrad, their only child, was born the following year. After years of acrimony and anger, with children caught in the middle, Peter and Russell divorced in 1952. Russell's fourth and final marriage, to Edith Finch, lasted serenely and congenially until Russell's death in 1970.

Yet Russell's marriages tell only part of tale. Having grown up in utter repression, Russell entered married life sexually and emotionally ignorant. His great awakening came in 1911, when he met Lady Ottoline Morrell, wife of Philip Morrell, a successful antiques dealer and member of Parliament. Their affair was marked by great passion on Russell's side and inevitable rancor when the relationship cooled. By then, Russell had taken up with Lady Constance Malleson, an actress whose stage name was Colette O'Neil. Russell befriended and may have had romantic relationships with several other women, including Evelyn Whitehead (the wife of his colleague Alfred North Whitehead), the writer Katherine Mansfield, and Vivienne Eliot (wife of the poet T. S. Eliot).

Russell and Dora Black began their relationship in 1919, two years before their marriage. In Dora, Russell found an intellectual partner. She was an activist and scholar, well traveled and independent. When Russell went to Russia in 1920, she followed, though their paths never crossed and they both returned home singly. Not so for Russell's next trip abroad. He and Dora spent the academic year 1920–21 in China (which he loved), during which time Russell lectured at the National University — and nearly died of pneumonia. Russell had traveled twice to the United States before the war; in 1924, he returned to lecture at universities, institutes, and clubs across the country. Subsequent tours from 1927 through 1934 helped Russell afford upkeep for his children.

Swept up by various radical philosophies of education, Russell and Dora began a school for children (including his own) in 1927. Russell and Dora tried to chart a middle ground between conventional schooling and the relative anarchy of "new schools." The experience left the Russells not only in debt, but disillusioned by the cruelty of the children. Their progressive pedagogy failed to stem bullying. One of the most vulnerable was their son John, who, like his sister Katherine, survived by retreating into a "shell."[102] Meanwhile, Russell's marriage to Dora was deteriorating; they separated in 1931. At the time, Dora was pregnant by her lover Griffin Barry, and Russell was cohabiting with "Peter" Spence. The school became Dora's, and she ran it until the war.

Russell had cast his lot far from the security (and orthodoxy) of academia for most of his life. In 1937, with the birth of his third child, Conrad, with Peter, he looked about for a permanent university position that would solve his financial problems. But none was forthcoming. With war on the horizon, Russell, Peter, and Conrad set sail for the United States in 1938 on the promise of a visiting professorship at the University of Chicago. For the next six years, Russell and his family moved from one position to another — or, in the notorious case of the City College of New York, which rescinded an offer when a mother complained that Russell was too radical to teach at her son's college, from one position to no position at all. With the declaration of war, Russell was stuck in the United States. His prolific writing did not cease, and he was a popular lecturer. But his reputation as a libertine and a sexual radical preceded him at every moment. In the end, his years in the United States did not solve his financial difficulties; they only served to intensify his longing to return to England. He did so, finally, in 1944, as the war waned, and he was finally able to secure transport for himself, Peter, and Conrad.[103]

Far from having been forgotten in England, Russell was received warmly. In early 1944, his old Cambridge college, Trinity, offered him a fellowship starting in the fall — healing the breach caused by his expulsion during World War I. Back in London, he became a popular lecturer on the BBC and soon a member of the highbrow quiz show, *The Brains Trust*. Then, in 1945, his *History of Western Philosophy* became a runaway best seller, at last ending his financial troubles at the ripe old age of seventy-three. In 1948, he became a folk hero of sorts. Flying from Norway to Sweden in a storm, his plane crashed and nineteen people drowned. But the seventy-six-year-old Russell swam through icy waters to a rescue boat, none the worse for wear. In 1949, he became truly respectable. He was made a member of the Order of Merit, Britain's highest honor to its intellectual and artistic elite. In 1950, while visiting Princeton again, he learned that he had won the Nobel Prize in Literature.

Russell certainly did not think of himself as having slowed down in his later years — nor did others. Hearing of Russell's financial straits in 1937, George Santayana immediately and generously sent the sixty-five-year-old Russell a yearly stipend. "Old and penniless" Russell might have been, but in Santayana's view, he was "still brimming with undimmed genius and suppressed immortal works."[104] Still, for all his energy, Russell's most important philosophic work had stopped in 1927, at about the same time as Einstein's most important scientic work. Russell's mathematical work lay in the distant past.

MATHEMATICS AND LOGIC

Russell's philosophical career can be divided into two parts: first logic, then philosophy of science. The split came in 1910. In that year, he and Whitehead completed the *Principia Mathematica*. It would be hard to overestimate the lasting importance of this work. It grew out of the profound conflict Russell felt as a young philosopher schooled, along with other Cambridge Apostles, in the idealism that was in vogue. Gone were the empiricists, including Russell's revered John Stuart Mill. Instead, he read Kant and Hegel and Berkeley. However beguiling, idealism was something thrust on Russell, a version of metaphysics that could not satisfy his longing for "truth." Absolute idealism locates reality within the mind. Only mental conditions and constructions are real. For Plato, the universal "ideals" constitute reality. Hegel distinguished between finite nature and infinite ideas, finding only the latter to be "real" — thus, his "absolute idealism," as opposed to Kantian "subjective idealism," which limits our knowledge to our mental impressions of the external world, which can (if it exists) be perceived only indirectly, through organizing "categories" inherent in the mind.

Russell, for all his exposure to idealism at Trinity, came away unconvinced. With G. E. Moore, for whom "common sense" was the rule, Russell abandoned absolute idealism, arguing that an objective world susceptible to analysis did, indeed, exist. When Russell taught Leibniz for a semester, he was thunderstruck by the latter's method of analysis. If language could be broken down to reveal its basic structure, so then could logical analysis become a tool for discovering the truth. And what better foundation for logic than mathematics?

The new century intervened with the First International Congress of Philosophy and the Second International Congress of Mathematics, held one after the other in Paris in the summer of 1900. The backdrop was the great Exposition Universelle. Whereas all of Paris was transformed by new architecture — the Grand and Petit Palais, the Eiffel Tower — Russell was transformed by the work of one man: Giuseppe Peano, the great mathematician of Turin. For months afterward, Russell read Peano's works and corresponded feverishly with him. Peano's system of symbolic notation was, Russell believed, extendable to the logic of relations. Fellow Apostle A. N. Whitehead joined Russell at his house in Fernhurst, and soon the two men committed to a collaboration on what would become the *Principia*.

In three volumes, written jointly over the course of ten years,

Russell and Whitehead, following Gottlob Frege, laid down the principles and elements of logicism. In short, logicism asserts that all mathematical truths can be stated in the form of logical truths and that mathematical proofs can be derived from logical proofs.

For the young Russell (he was thirty), mathematics was a haven from his increasingly unhappy and complicated private life. He had fallen out of love with Alys and was embroiled, romantically though probably not sexually, with Whitehead's wife, Evelyn, who suffered terrible pain from angina. Some forty years later, Russell was to remark caustically that Gödel was mired in Platonism. Yet he, too, felt drawn towards the "enchanted region" of mathematics, where "in thinking about it we become Gods."[105] In a letter to his friend Gilbert Murray, he pronounced mathematics and philosophy to be concerned with "ideal and eternal objects."[106]

As he and Whitehead worked through the intensely technical matters of the *Principia,* Russell must have found those "ideal and eternal objects" increasingly remote. By its nature, the *Principia* led Russell face-to-face with paradox, the inevitable spanner in the mechanics of logic. In answer to his own famous paradox (To what class does the class of all things which do not belong to themselves belong?), Russell wrote "On Denoting," delivered in 1905. He was able, paradoxically, to construct a "no-class theory," taking both classes and numbers out of the realm of the ideal.[107] What was left — propositions — still carried weight as Platonic "truths," but soon, this "haven of peace" disappeared, to be replaced by the doubt more suited to his empiricist roots.

It took seven years of extraordinarily intense and exhausting work to complete *Principia Mathematica.* After that, Russell ceased, for all practical purposes, to do highly technical and demanding work on logic. He was thirty-eight years old. He had come to hate the shuttered concentration that logic demanded. Writing the *Principia* was like juggling several dozen balls at once for years on end. In a logical system, he said, writing to his longtime friend

Lucy Donnelly, "one mistake will vitiate everything." The toll, he acknowledged, was tremendous. He described at length, and in dramatic fashion, the "sheer effort of will" necessary for such work:

> Abstract work, if one wishes to do it well, must be allowed to destroy one's humanity; one raises a monument which is at the same time a tomb, in which, involuntarily, one slowly inters oneself.[108]

And, indeed, these words were written in 1902, when Russell had finished his precursor work, *The Principles of Mathematics*. In 1910, having finished the much longer and grander *Principia*, he was "somewhat at loose ends. The feeling was delightful, but bewildering, like coming out of prison."[109] He never went back in.

In *My Philosophical Development*, written half a century later, Russell thinks back upon his devotion to a nonhuman, idealist mathematics. As the contradictions mounted, Russell lost that devotion. He came to accept Wittgenstein's dismissal of mathematics as "tautologies." In the face of "young men embarking in troop trains to be slaughtered on the Somme because generals were stupid," mathematics and the "world of abstraction" were for all intents and purposes lost. Perhaps his disappointment lay in his character. As one critic has it, "as one reflects on Russell's philosophical career, it appears that behind this thirst for certainty there lurked an even deeper craving for disillusionment."[110] Still, once the godson of John Stuart Mill had completed the monumental *Principia*, his world changed.

From mathematics, he turned to philosophy. Although (perhaps because) Russell wrote some seventy books and hundreds of essays on philosophy and philosophical topics, his own philosophy is difficult to summarize. Like Einstein, he hated disarray in the foundations of knowledge. His major philosophical work examines the premises and beliefs undergirding logic and science. The titles of his important works illuminate his grand scope: *The*

Analysis of Matter, Human Knowledge: Its Scope and Limits. But his was truly an experimental and question-driven philosophy. He was always ready to try a new approach to find solutions. He often revised his views, but never thought this a failing. In a way, he modeled his philosophic approach on the piecemeal and provisional approach of physics. He never built a grand system. Instead, he inspired the modern movement known as "analytic philosophy." Like Russell, his philosophical heirs, Wittgenstein and the logical positivists, were better at dissecting than building.

Once Russell turned away from logic, where he had made his true mark, he looked toward physics as the sole arbiter of certain knowledge. Modern physics, he thought, had the best chance of being true about the external world. What was left over, the empirical world that we know through our senses, yields information quite different from the truths of physics. Most of Russell's philosophical career was spent pondering these two paths towards truth. Logic, his first passion, was no longer the high-road, only a tool.

Still, Russell was drawn back into his early world of mathematical logic from time to time. In *The Philosophy of Bertrand Russell,* a collection of essays in Russell's honor, several contributors revisited the *Principia* and Russell's place in the history of mathematical logic. Russell dutifully commented on all contributions, save one, written by the agonizingly exacting Kurt Gödel and submitted months late.[111]

UTOPIAN ENEMY AGENT

Russell was forty-two when war was declared in August 1914. He was already seen as one of the world's important logicians and philosophers, having been elected a Fellow of the Royal Society in 1908 at only thirty-six and offered a teaching position at Harvard in early 1914. Up to then, his life had been that of a scholar and teacher. But he was restless and uncertain. His love affair with Ottoline Morrell began that year. It was as if, once he left mathematical logic behind, his sexual passions exploded. Yet his intellectual engine never stopped. As always, he turned out a prodigious number of books, articles, reviews, lectures, and letters. As for the war, it so changed Russell that he later thought of himself as a Faust figure who, on that fateful August day, met his Mephistopheles.[112]

The war swept him into new roles. He became a pacifist, a war resister, and a man of action. None of this could have been predicted. Up to then, he had supported Britain's colonial wars — the Boer War, for instance, had been a necessary adjunct to the spreading of "civilized government."[113] But he could see no sane purpose in the European war, and said so with increasing bite and fury. Once he was committed, his courage and defiance never wavered, though he was berated as a traitor to his country and class. He was, after all, the Honorable Bertrand Russell, grandson of a prime minister, son of a lord, and brother of an earl. He lost close friends. The otherwise cherubic Alfred North Whitehead, his collaborator on the *Principia,* caught war fever and could not abide Russell's lack of patriotism. His old friend, the Greek scholar Gilbert Murray, denounced Russell as "pro-German" in print. Russell understandably took to decrying the "bloodthirstiness of professors."[114]

For almost four years, his life became a marathon of political maneuvering, writing, and speaking. At the start of the conflict, he helped found the Union of Democratic Control, an antiwar movement. Russell suddenly blossomed as a mover and shaker. The UDC boasted such influential figures as the future prime minister Ramsay MacDonald; the peace activist and future Nobelist Norman Angell; the journalist and tireless campaigner against King Leopold's Congo, E. D. Morel; the writers Leonard Woolf and Lytton Strachey; and Russell's old Cambridge friends Lowes Dickinson and Charles Trevelyan. Russell soon dominated the movement. As an observer noted, "No resisting the force of his ruthless dissection of motive; no reply possible to the caustic comments he would emit in his high squeaky voice."[115]

By the summer of 1915, after only one year of war, over a

quarter million English soldiers had been killed or wounded, all of them volunteers. Britain had never contemplated a conscription law. In the face of this mass slaughter, it did so, and in January 1916 a law was passed requiring all males between eighteen and forty-one to register for military service. The UDC voted not to oppose conscription, and Russell quit. He had consistently reproved the UDC for having "no intensity of will."[116] He promptly joined the more radical No-Conscription Fellowship (NCF), which had supported conscientious objectors since the war's outset.[117]

These new experiences and emotions liberated Russell as never before. In 1916, he gave lectures on a new theory of society based on "creative" and "instinctual" alternatives to the destructiveness of war (published as *The Principles of Social Reconstruction* in 1916). It was the first of many briskly rational, quasi-utopian proposals that he launched periodically throughout his long life. He gave stump speeches and orated at rallies. He visited conscientious objectors in prison and lobbied tirelessly on their behalf, salvaging their mental health and perhaps even their lives in the face of ferocious governmental hostility.[118]

At times, his antiwar fervor bordered on the obsessive. "It is a real ferment," he wrote of the no-conscription movement, "like the beginning of a new religion." Religion indeed: "I rather envy the men they persecute. It is maddening not to be liable."[119] His young comrades, admirable as they were, lacked "the thirst after perfection — they see the way out of Hell but not the way into Heaven." Yet "they will joyfully become martyrs."[120] In upswing moods, he declared that "I want actually to *change* people's thoughts. Power over people's minds is the main personal desire of my life."[121] As one UDC member shrewdly noted, Russell

> had a dynamo within that was too powerful for his own comfort and far too powerful for that of others: inevitably, he first swallowed admirers and then, with what they felt a heartless cruelty, spewed them out.[122]

Russell quickly arrived on the government's list of trouble-makers. Fearful that he might travel to the United States and foment opposition to the British effort, the authorities sought a reason to refuse him a passport. In 1916, his wish for martyrdom nearly came true. A No-Conscription member was sentenced to two years at hard labor, and six others were then sent to prison for circulating an anonymous leaflet protesting the case. Russell publicly admitted writing the leaflet and was arrested and fined £100. When Russell refused to pay, the authorities impounded his books and furniture from Trinity.[123] The conviction allowed the government to revoke his passport, forestalling a showdown. Trinity College, his alma mater, quickly used the conviction as an excuse to remove him from his lectureship. Whitehead and others protested. But Russell was (at least outwardly) euphoric at the news. "I no longer have the feeling of powers unrealised within me, which used to be a perpetual torture. . . . I have no inward discords anymore."[124]

By now the government spied on Russell, absurdly, as an "enemy agent."[125] He was banned from restricted areas, lest he signal enemy ships — an absurd idea, though a convenient cover for stopping Russell from lecturing to and encouraging conscientious objectors.[126] Russell tried another tack, writing a letter to President Wilson urging him to force Europe to the peace table. The letter managed to slip by the censors of the Foreign Office. Wilson ignored it, but the letter (and details of its secret journey) was printed in full by The New York Times.

Russell continued his antiwar efforts. When the Russian Provisional Government put forth a peace offer, he was ecstatic. With great fervor, he threw his support behind the revolution and its British admirers. In July 1917, however, a meeting of revolutionary sympathizers at Hackney disintegrated into violence, leaving Russell shocked and disheartened. He returned to his philosophical work, spending the fall and winter writing and lecturing on logical atomism.

Ironically, just as Russell had become disillusioned with the efficacy of protest, he was arrested in 1918 for "insulting an ally" — the United States, which had entered the war. The alleged crime — he had written a short article advocating peace with Germany — was the pretext for a harsh sentence: six months at hard labor in the so-called "second division." The sentence was not to be taken lightly. Long stretches in the second division had left his colleagues Clifford Allen and E. D. Morel physically devastated, and men could be crippled during such a sentence. Friends, including Gilbert Murray, brought pressure on the government to shift Russell to the "first division." [127] At his appeal, the magistrate, citing Russell's contributions in logic and philosophy, acceded. Russell served his six months in the relative comfort of the "first division." Because he could pay, he had a large separate room, with meals brought in from outside, a servant to clean the "cell," daily delivery of the *Times,* and a well-stocked library of chosen books. Russell compared it to "life on an Ocean Liner."[128] Visitors came three times a week. In such enforced but tolerable isolation, the exhausted Russell revived and soon wrote two books, an *Introduction to Mathematical Logic* and a draft of *The Analysis of Mind.*[129]

Even with this seriocomic finale, Russell's career as a war protester makes Einstein's antiwar efforts pale by comparison. The anti-war movement energized Russell and propelled him forward. He saw in the future "infinite possibilities." It would be hard to guess from this excited language that he meant teaching philosophy to "working-men who are hungry for intellectual food. . . . Think of building up a new free education not under the State! . . . I could give heart & brain & life to that."[130]

In the Wilderness: Between the Wars

The day World War I ended, Russell in victorious London was depressed: Millions had been pointlessly slaughtered, but people were wildly celebrating in the streets. Russell had spent the war years in feverish political activity. Almost fifty years would pass before he plunged again into antiwar protests, against the nuclear bomb and the Vietnam War.

During the last half of his life — from 1920 on — Russell's affection for his country grew in tandem with his popularity. He became the plain-speaking oracle, the dauntless opponent of injustice and folly, the philosopher with a gift for connecting to the common people. Russell the philosophical popularizer blossomed after the First World War. Had he not been radicalized by that war, he would likely have returned to teaching philosophy and logic, his works known only to an inner circle of specialists. The oracle and gadfly would have been stillborn. As it was, he never returned to a full-time academic career. He became, instead, a freelance writer, an educational innovator, and a prophet of social change.

In crucial part, Russell's popularity stemmed from his passionate belief in the usefulness of philosophy. Unlike many of his fellow academics, Russell had taken up philosophy to find consolation and meaning in life. For him it was no academic exercise. "I wish to understand the hearts of men," he wrote in his *Autobiography*. This desire may have led him to abstruse mathematics, but it was nonetheless ordinary and human. On his journey, he experienced a "failure" that was yet a "victory. . . . I may have conceived theoretical truth wrongly, but I was not wrong in thinking that there is such a thing, and that it deserves our allegiance."[131] Russell has been consigned to history, rather than philosophy, by a modern tradition that prefers the technical to the metaphysical. Yet, notes Frank McLynn,

> Russell was that rare bird, a professional philosopher who actually tried to answer the questions that ordinary people naively imagine can be answered by philosophy. He was in fact a "philosopher" in a sense that would be recognised by the man in the pub. This was why he, alone of his breed, could move between the worlds of Whitehead and Wittgenstein and those of Conrad and Lawrence.[132]

And, as Michael Foot writes, "[a] particular, persistent reason" for his "appeal, throughout his ninety-odd years, especially to the young, was the trouble he took to write plain English."[133] In recent years, it had become not only fashionable but occupationally imperative for academic philosophers to write for other academic philosophers rather than for a general reader. Despite his ability to write (with the more mathematically adept Whitehead) the *Principia Mathematica,* Russell was no technician. Alan Wood has described him as "a philosopher without a philosophy. The same point might be made by saying that he is a philosopher of all the philosophies."[134] In later years, he came to believe in philosophy writ large — in other words, philosophy that is concerned with "matters of interest to the general educated public, and loses much of its value if only a few professionals can understand what is said."[135]

The aristocratic Russell had one thing in common with the populace for whom he wrote: He was perennially short of money. He was fifty when John, his first son, was born. Delighted as he was with the novel sensations of parenthood, he faced the "inescapable responsibility" of providing financial support. To that end, he churned out potboilers on sex, marriage, and divorce; on conquering happiness and praising idleness; on atoms and relativity. He became a regular columnist for the American Hearst newspapers. He lectured across the United States several times in the 1920s and 1930s, and in his seventies was a popular voice on the BBC.

In 1938, badly needing a steady income, he did try to return to teaching, but not in England. He moved his family to the United States to find a suitable university position. A series of small fiascos ended in two big ones. He taught at Chicago (they wouldn't keep him) and Los Angeles (where he quarreled with the chancellor), and then was appointed in 1940 to teach philosophy at the City College of New York. There, the political and religious establishments blocked the appointment, accusing him of immorality, incompetence, degeneracy, godlessness, anti-Americanism. His works were damned (in the words of one lawyer) as "lecherous, venerous, lustful, erotomaniac, aphrodisiac, irreverent, narrow-minded, un-truthful and bereft of moral fiber."[136] Einstein, Whitehead, John Dewey, and even Charlie Chaplin rose to his defense, but in vain. After months of fighting, with hate mail pouring in, Russell's position was simply eliminated. He lectured at Harvard (an engagement that predated the City College debacle), but thereafter American universities shunned him. Compounding his dire financial straits was his escalating disdain for America. He was homesick for England, which had survived the Battle of Britain but still faced great danger.

Russell was rescued by an eccentric millionaire in Philadelphia. Dr. Albert Barnes, a chemist, had spent his fortune (made on the drug Argyrol) amassing French Post-Impressionist paintings. His private museum housed hundreds of Picassos, Cézannes, Matisses, and Van Goghs. It was open only to a select few, those whose taste suited him. In late 1940, on the recommendation of John Dewey, Barnes offered Russell a handsome salary to give popular lectures on philosophy at the museum. Russell began what was to be a five-year term in January 1941. At first, Barnes was enthusiastic. Within a few months, however, he began meddling in Russell's classes. There were quarrels. Barnes's ego was further bruised by Russell's wife, Peter, whom he deemed "imperious" and banned from the museum. Barnes fired Russell a few days after Christmas 1942. Russell sued for breach of contract and won, but was forced to wait months for payment. Beginning in 1943, he tried to find transport back to England, but it was the spring of 1944 before he, his wife, and their young son Conrad finally embarked. Meanwhile, he kept busy writing his *History of Western Philosophy*.

As always, he planned ahead. What would become his last work of philosophy, *Human Knowledge: Its Scope and Limits,* was in its planning stages. Throughout October 1943, Russell delivered a series of five lectures on successive Fridays at Bryn Mawr College. The lectures were received by an enthusiastic audience who braved "torrential rain." Their titles are notable to us, for they suggest what was on Russell's mind at the time: (1) "Limitations of Deductive Logic," (2) "Probable Inference in Practice," (3) "Physics and Knowledge," (4) "Perception and Causality," and (5) "Induction and Analogy."[137]

Human Knowledge was to be, in the words of Ray Monk, Russell's "last major philosophical work."[138] His purpose was "to examine the relation between individual experience and the general body of scientific knowledge" — in sum, the age-old dialectic between the concrete and the abstract, applied in particular to the world of science. Philosophy flourishes as an adjunct of science, especially physics. The problem is to find the link between what we see and what is there (always the problem in epistemology), or, in other words, the common world around us and the world described by science. It is telling that "individual experience" comes first in his thesis. For, again, Russell held always to the world of experience, however desirous he was of an overarching certainty. Yet, as the philosopher A. C. Grayling remarks, "he was ... critical of certain forms of empiricism" because a focus on "sensory experience,"[139] the very definition of empiricism, cannot account for scientific knowledge. Thus, *Human Knowledge* takes up the problem of "non-demonstrative inference," the primary method by which science works, and the difficulty of finding structures to ensure truth-finding in science.

Pondering these questions, in late 1943 or early 1944, Russell rented a lakeside house near Princeton. There, once a week, he walked in the bitter cold to 112 Mercer Street and chatted with Einstein, Gödel, and Pauli.

GÖDEL: GHOST OF GENIUS

Einstein's closest friend at Princeton was Kurt Gödel. The wonder is that they were friends at all, so different were they in temperament and style. Einstein was twice Gödel's age. He loved jokes and laughter. He was generous, down-to-earth, and the epitome of sanity. Gödel was distrustful of people's motives, a hypochondriac, often depressed and paranoid. In the end, he starved himself to death, convinced that his doctors were trying to poison him. One cannot imagine Gödel enduring what Einstein took in stride — wearing an Indian war bonnet for photographers, chatting with Charlie Chaplin or Winston Churchill, trading cookies with a neighbor's child. Einstein loved Bach and Mozart. Gödel said that Bach made him "nervous" (his taste ran to "O Mein Papa" and "The Wheel of Fortune").[140] Einstein played the bohemian. When he lived alone in Berlin, he cooked soup and eggs all together in the same pan to save time. Gödel was a thorough and contented bourgeois, living snugly with his wife in a Princeton bungalow full of kitsch. When his wife set a pink flamingo on the lawn, Gödel thought it "terribly cute."[141] Einstein felt compelled to fight injustice, though it cost time and energy away from physics. Gödel, though he had fled Nazi Vienna, never so much as glanced up from his equations.

How could their friendship thrive? Clearly, it did. Late in his life, Einstein told a friend that when he felt old and his own work no longer meant much, he came to the Institute mostly for the privilege of walking home with Gödel.[142]

The word "privilege" salutes the younger man with Einstein's typical generosity. It also hints at Einstein's isolation from the greater scientific community. Gödel helped fill that void in Einstein later years. Gödel believed that Einstein liked him because he was willing to argue. But intellectual stimulation is hardly the sole basis for a close friendship. The intriguing question is: Why did Einstein *enjoy* Gödel's company?

Some light is shed by Ernst Straus, Einstein's mathematical assistant in 1944. He wrote that although the two men differed "in almost every personal way," Gödel "in some ways strangely resembled [Einstein] most. . . . They shared a fundamental quality: both went directly and wholeheartedly to the questions at the very center of things."[143] They were also, as Palle Yourgrau notes, equally "unapproachable" because of the "sheer size of their reputations."[144] They shared a common language, German, and common interest in each other's fields — Einstein had finally awakened to the importance of mathematics, and Gödel had once dabbled in physics. They were strangers together, philosopher-kings in the brave new world of technophysics. When it came to understanding intellectual questions — indeed, to tackling them in the first place — Gödel had the requisite audacity and inner freedom.

Gödel's towering reputation was built on two theorems, collectively known as "incompleteness." The audacity of incompleteness is best illustrated by a story about John von Neumann, a mathematician of prodigious output, far exceeding Gödel's in number and range. In 1930, von Neumann — only three years older than Gödel — was pondering the same questions as Gödel. But he failed where Gödel succeeded. For in proving his theorems, Gödel disproved David Hilbert's mathematical formalism. Hilbert was not only the greatest mathematician of the time, but von Neumann's mentor as well. Had von Neumann ventured to conceive of incompleteness, he would have had to imagine the towering Hilbert capable of error.[144] Gödel — like Einstein — was never awed by eminence, and never had nor wanted a mentor. In such intellectual matters, Gödel was as self-contained and self-confident as Einstein.

Outside mathematics and philosophy, Gödel was helpless, even infantile. He needed care and attentiveness and, at times, nursing. His wife, Adele, mothered him from the beginning. During their engagement, he was terrified of being poisoned. She loyally tasted all his food to reassure him and patiently fed him "spoonful by spoonful" to build up his weight.[146] He found in Einstein another protector. Gödel's frailty evoked in Einstein both sympathy and tenderness. Perhaps Einstein, nearly twice Gödel's age, saw in Gödel something resembling a son, brilliant and troubled. Einstein's youngest boy, Eduard, was schizophrenic and confined to a Swiss sanatorium.

Few would mistake either Einstein or Russell for a mere professor: Einstein looked like a wild-haired sage; Russell might have stepped out of the House of Lords. But Gödel looked the part, unworldly and abstracted. Nothing in his life corresponds to Einstein's commitment to Israel or Russell's antiwar crusades. Great minds are not necessarily great men. Einstein and Russell lived their greatness in the public eye. Gödel, their intellectual peer, was otherwise fragile and somewhat diminished.

He was born in 1906 in Brunn ("Brno" in Czech), a province of the Austro-Hungarian Empire, and attended schools there. Then he moved to the University of Vienna and took his doctorate in mathematics in 1930, at age twenty-four. That same year he found the breakthrough to the epochal proof that any system that is logical and consistent must be incomplete, which was published the next year. His life seemed ordinary in many respects. He took classes, went on vacations with his family, spent time in Vienna cafés, and had an eye for the girls. His father died in 1929, and his mother moved in with Gödel and his brother in Vienna. He had severe rheumatism at age eight, a nervous attack of some kind at five, and several depressive or psychotic episodes in his late twenties.

Yet it is hard to bring the young (or later) Gödel into focus. He never showed much of himself to anyone. Contemporaries described him as quite interested in what others said but saying little himself, speaking precisely but very briefly when he did, and rarely on topics outside mathematics.[147] Alert but withdrawn — he never changed; his watchful photographs suggest this. Even his publishing record can be described as withdrawn: He wrote much but published very little, because he disliked exposing himself to controversy or criticism. Cautious about every word destined for public scrutiny, he could delay promised material for years. His greatest work, the first incompleteness theorem, was announced at a mathematics conference with such modesty that it almost escaped unheard.

Still, glimpses of the unexpected emerge. Gödel's marriage was such a case. He met his future wife, Adele, in 1927, when he was twenty-one. She was a nightclub dancer, married at the time to a Viennese photographer, and eight years his senior. Marrying Adele was a gamble. If Gödel were to have a chance at an academic life, he would be expected to uphold high social standards. The title "Professor" carried with it civil service status, and thus much scrutiny. (As the hapless Herr Professor Rath discovers in the 1928 film *Blue Angel,* marrying a dancer invites merciless scorn and degradation.) Gödel's mother and brother were opposed to the marriage. The Gödel family was well-to-do, solidly middle class, well educated; Adele would not fit in. She was distinctly lower-class. She was uneducated and Catholic, with a port-wine stain on her face, and later developed a habit of bullying Gödel.[148]

The otherwise devoted son and sober academic Gödel defied family and propriety when, after several years of secret engagement, he married Adele. Both families were represented at the private ceremony, but Gödel's brother had never met Adele before, and none of Gödel's friends were notified of the wedding date.[149] Man and mouse at once: He seems to have wanted a wife who would protect and mother him, yet he acted boldly and cunningly enough in making sure he married her.

Adele was hardly a typical Princeton wife. Gödel's closest friend apart from Einstein, the Princeton economist Oscar Morgenstern, minced no words: She was a "Viennese washerwoman type: garrulous, uncultured, strong-willed" whose "astonishing bad taste" in décor he deplored. His wife Dorothy was "roused to indignation" by Adele's treatment of the frail and retiring Gödel, noting, among other trespasses, that Adele smoked.[150]

Still, Adele was a good choice for Gödel. She had the outgoing toughness and directness that he lacked. (After Germany annexed Austria, Nazi gangs roamed the streets attacking those who seemed to belong to the wrong side. In 1939, walking with Adele near the university, Gödel was assaulted and his glasses knocked off. Adele counterattacked with her umbrella and saved him.) His great discoveries in logic came before the marriage, but Adele very likely kept his later career going, especially as he became more hypochondriac and fearful. She was adaptable as few wives might be. In Princeton, Gödel believed that gases from the furnace might poison him, so the heat was turned off — even during winter.[151]

Their home life otherwise revolved around the quiet middle-class pleasures of "house and garden, food and digestion, household helpers, weekly accounts, summer vacations . . . families and relatives, anniversaries, birthdays," and the like.[152] Gödel arranged his life to avoid disturbing his concentration.

The marriage was probably happier on Gödel's side than Adele's. She spoke English badly, had little interest in intellectual pursuits, and found Princeton stifling. When Gödel refused to move to a livelier city, she escaped by traveling frequently to Europe. Gödel was a recluse at best. He was uncomfortable in social situations. If she could be a shrew — at times, Gödel said, hysterical — they suited each other. Despite all their differences, Gödel and Adele remained married until his death.

Gödel came of age in turbulent Vienna, center of a world in chaos. After its defeat in World War I, the Austro-Hungarian Empire was dismantled. Economic pressures mounted. Fascists, socialists, and Nazis fought in the streets. In early 1933, civil war broke out. Chancellor Dollfuss abolished the parliament, set up a dictatorship, and suppressed socialists and Nazis alike by force. In July 1934, Dollfuss himself was assassinated by the Nazis. Weak governments followed. Hitler invaded in 1938 and turned Austria into a province of Greater Germany.

Gödel, meanwhile, graduated from the University of Vienna in 1930 and began the difficult task of finding work among the rigid and closed hierarchies of Austrian and German universities. His incompleteness theorems of 1931 — though as epochal as relativity — seemed to many mathematicians to be off in the remote margins where logic and mathematics met. So, highly esteemed by a few but otherwise undistinguished and unconnected, Gödel searched for a position as a Dozent, the lowest rung on the academic ladder. Professors had tenure and received their salaries regardless of whether many or none took their courses; a Dozent had no tenure and no salary, and collected fees only according to how many students enrolled. In summer 1933, Gödel taught his first course at the University of Vienna. It would be twenty years, perhaps, before he could expect to be named professor.

However, his prospects improved drastically even before he began his summer course. The young mathematician John von Neumann had been present when Gödel presented his incompleteness theorems. Unlike many others, von Neumann immediately grasped the significance, and he began spreading the word in Princeton, where he had taught since 1930. Oswald Veblen, the senior mathematician at the newly formed Institute for Advanced Study (IAS), invited Gödel to visit the Institute, starting in the fall of 1933. His year went well and included a trip to Cambridge to lecture at the Mathematical Association of America. He returned to Europe in May and was in Vienna by early June. The summer brought more political chaos and a subtle escalation of Nazi influence. Although scheduled to return to the IAS in the fall of 1934, Gödel had a nervous breakdown, which he hid from Veblen (claiming an infected tooth) and spent several weeks in a sanatorium. One year later, he returned to Princeton. The voyage must have been stimulating, for he sailed with Wolfgang Pauli and the mathematician Paul Bernays, also en route to the IAS. But within two months, Gödel suffered a recurrence of depression and returned to Vienna, with Veblen's assurance that he could return at any time. Gödel's lifelong mental instability seemed to have taken shape. He spent much of the remaining year reading up on toxicology and psychiatry. His marriage in 1937 calmed him, and he was able to return to Princeton in the fall of 1938.

Meanwhile, his constant leaves of absence from the University of Vienna and his inattention to bureaucratic requirements came back to haunt Gödel. He had neglected to request a leave in 1938 until well after he was ensconced in Princeton. Having caught the attention of the Ministry of Education, he returned in 1939 to a Vienna dramatically changed under the Nazi takeover. His Dozent appointment was on the verge of expiring.[153] Only a "New Order Dozent" position — bestowed at the pleasure of the Nazi regime — was available. Gödel filled out an application. A Nazi bureaucrat with the inimitable title of *Dozentenbundführer* (Leader of the Association of Dozents)[154] reported that Gödel's doctorate had been directed by a Jewish professor (Hans Hahn) and that Gödel "always traveled in Jewish-liberal circles."[155] However, there was no record of Gödel's having disparaged Nazism. The application lingered in officialdom. Meanwhile, Gödel was examined for military duty and, astonishingly, found fit to serve. Facing the unthinkable, Gödel acted quickly, negotiating quietly for a visa to the United States. Despite bureaucratic tangles, Flexner at the IAS was able to wangle a special visa for him. Only after the incident with the Nazi thugs, however, did Gödel decide to leave Austria. To obtain transit visas, he and Adele were forced to take a circuitous route through the USSR (then a Nazi ally) to Japan, where they set sail for California. Within days of landing, the Gödels arrived safely in Princeton.

In retrospect, Gödel's behavior in 1938 and 1939 is puzzling and disturbing. He was in Vienna on March 12, 1938, when Germany invaded Austria. The next day, Hitler spoke to a hundred thousand people in the Heldenplatz and decreed Anschluss, annexing Austria to Germany. Almost immediately, Austrian anti-Semitism burst out violently. Persecution of the Jews was savage. Crowds watched gleefully as Jewish doctors, businessmen, and well-dressed women were forced to scrub the sidewalks of Vienna with toothbrushes until they were clean of anti-Nazi slogans. The third largest Jewish community in the world — and the most sophisticated — was suddenly stripped of dignity, position, and employment. Stores, homes, and synagogues were broken into and looted; many thousands of Jews were imprisoned or shipped to concentration camps.

Even though he had been in the United States during some of the worst atrocities, the reclusive Gödel could not have been unaware of what was happening. Jewish (and liberal non-Jewish) academics were among the first targets. Colleagues, friends, and neighbors disappeared overnight to exile, prison, or death. He was surrounded in America by refugees and well-informed opponents of Nazism.

The answer lies not in his political inclinations, but in his psyche. He was neither anti-Semitic nor pro-Nazi. Rather, he was excessively detached. When he had lunch with the newly arrived Austrian-Jewish mathematician Gustav Bergmann, driven into exile in 1938, Gödel stunned Bergmann by asking, "And what brings you to America, Herr Bergmann?"[156] The mathematician Karl Menger, an old friend from Vienna, recalled that he never uttered a word to Gödel about the horrors unfolding in Europe simply because Gödel seemed unconcerned — except for the threat to his Dozent position.[157]

One can only wonder how Gödel's friendship with Einstein survived such a striking difference in temperament and outlook. Nothing was further from Gödel's self-absorption than Einstein's engagement with the world. Einstein never hesitated to break a rule or disregard legalities to do right. Gödel's obsession with legalities struck everyone who knew him. To be legalistic is to cleave to the letter of the law — the rules — rather than to the spirit, and thus to a narrowed view of what is at stake. Adhering to the letter was hardly appropriate in a world ruled by Nazis.

Gödel's obsessive legalism, the consequence of an extreme rationalism, was famous. Whereas a reasonable person accepts the muddle of life's confusion, imperfection, and accidents, the extreme rationalist seeks to banish dilemmas by so precisely defining problems that the solution seems simple. (The paranoid is perhaps the only perfect rationalist: Every seemingly random act, gesture, look, or word is part of a logical pattern; no stray threads, accidents, coincidences, luck, or flukes can exist.[158] Gödel, in the grip of such a view, finally starved himself to death, convinced he was saving himself from untrustworthy doctors.[159]) Sometimes, Gödel's attempts to rationalize the world were comic. In 1947, he had to travel to Trenton for an interview regarding his application for citizenship. Einstein and the economist Oscar Morgenstern went along. On the way, Gödel remarked that he had studied the Constitution very closely, expecting to be drilled on its contents, and had discovered a fatal flaw in its logic. Alarmed, Einstein and Morgenstern cast about for ways to distract Gödel, lest he bungle his interview by lecturing on the Constitution's shortcomings. Einstein told joke after joke about whatever came to mind. When at last Gödel's turn came, the justice asked an innocuous question: "Do you think a dictatorship like that in Germany could ever arise in the United States?" Gödel saw his opening. "I know how that could happen," he said, and began to explain that fatal flaw. The judge, who knew Einstein, quickly changed the subject — and approved Gödel's citizenship.

When Gödel returned to Princeton in 1940, he was given only temporary appointments, to be renewed each year — this despite his renown. Only in 1946 was he given a permanent appointment — still as a visitor, not as one of the permanent faculty. In 1953, at age forty-seven, he was finally made a professor — and even that came after he had been awarded honorary degrees by Yale and Harvard and shared the first Einstein Medal in 1951.

Why had it taken so long for a professorship? For his part, von Neumann thought it absurd: "How can any of us be called professor when Gödel is not?"[160] Doubtless the delay was due to considerable apprehension on the part of several faculty members worried about Gödel's mental health. Others were irritated by his legalistic and interminable arguments during committee meetings and his habit of phoning at all hours of the night to talk about trivial matters. Eventually, of course, he was promoted. Gödel himself seemed indifferent and was never heard to complain about his treatment.

Gödel can scarcely be summed up by his private aberrations. Mathematics is the most highly rationalized form of thought, and Gödel was one of the great mathematicians. What is most intriguing is that he discovered contradiction and finitude in the heart of mathematics. The incompleteness theorems for which he is famous prove that no mathematical system can be both consistent and complete. His proofs are elegant and irrefutable.

Within this hyperrationalist mathematician lived a seemingly obverse personality.[161] He believed in ghosts, for instance. Our instinct is to ascribe such a belief to psychological problems. Except for frightened children, who takes ghosts seriously? Yet Gödel's university mentor, Hans Hahn, and the philosopher Rudolf Carnap frequented mediums. How do we reconcile their beliefs with a rational world view? Gödel never explained how he came to think that ghosts exist. Did he himself see them? Or read about them in fairytales? Or dream them? Yet are ghosts so different, for instance, from the Platonic idea of numbers? Such numbers are as puzzling as ghosts — they exist only as mental entities, but they are nonetheless real, not invented but discovered, in the same sense that astronomers discover stars and zoologists discover different species of life. Hence, reason is not mere finger play in the void; the objects of reason are inscribed in the eternal order of things.

Now, consider a "fact" about ghosts: They *linger*. If they didn't, would we ever speak of them? But why do they linger? Because (tradition says) they must abide until some higher task is completed: Their sins must be exorcised or their earthly longings abated. Not surprisingly, Gödel also believed in reincarnation, and he thought through some of the implications with philosophic rigor. In 1961, Gödel's mother asked whether he believed in a next world. Gödel said he did. Since the universe and life seemed regularly ordered, there might well be another life in another world:

> What would be the point of bringing forth an essence (the human being) that has so wide a range of possible (individual) developments and changes in their relations, but is never allowed to realize one thousandth of them? That would be like laying the foundations of a house with the greatest trouble and expenditure, and then letting the whole thing perish again.[162]

Thus, the soul (or human essence) proceeds from this world to the next and is reincarnated or reembodied so that it may continue its improvement.

If reincarnation is possible, "time" with its familiar panoply of past, present, and future becomes more mysterious than ever. In 1949, in a volume honoring Einstein, Gödel offered a new twist on Einstein's relativity: rotating universes whose effect is not only time travel, but the disappearance of time itself. Gödel's work on time is the subject of Palle Yourgrau's *A World Without Time: The Forgotten Legacy of Gödel and Einstein*.[163] Yourgrau is the first scholar to examine Gödel's philosophizing on time in the depth it deserves.

Gödel's famous incompleteness theorem is ultimately a limiting case for mathematics, demonstrating that mathematics cannot complete its proofs. As we shall see, the confidence that mathematics is a complete system was one of the high water marks of optimistic rationalism in the twentieth century. Incompleteness proved this confidence misplaced: No enterprise of human thought can escape limitation, not even the crystalline structure of mathematics. It might have seemed to Russell that Gödel was mired in Platonism, but on some very fundamental level, it was a Platonism tempered by reasonable doubt.[164]

PAULI: THE DEVIL'S ADVOCATE

On the evening of May 26, 1955, Wolfgang Pauli arrived at the Zurich Physical Society ready to lecture on Einstein. It was the fiftieth anniversary of the special theory of relativity. He expected to see among the many invited guests his two assistants, David Speiser and Armin Thellung. Both were missing.

So, too, were Ralph de Laer Kronig, in town on his annual trek back to Zurich, and Res Jost, an old friend and associate professor at ETH (Eidgenössische Technische Hochschule, the former Zurich Polytechnic School), Pauli's home university.

Kronig and Pauli were old friends, despite their inauspicious first meeting. In 1924, Kronig, a freshly minted Columbia Ph.D., had gone to the ever-skeptical Pauli, known as the "scourge of God," with an idea about electron spin. Pauli pronounced the idea "amusing" ("Das is ja ein ganz witziger Einfall") but unrealistic. He later regretted his words. Discouraged, Kronig gave up. In the months to come, two students of Paul Ehrenfest published very similar ideas on spin, and finally Pauli took notice. Building on these ideas, Pauli proposed a fourth quantum number with two values, one each for clockwise and for counterclockwise spin. Ehrenfest's students were ultimately awarded the Max Planck medal for their discovery. Kronig, quick to forgive, went on to a distinguished career, first at Zurich, where he was hired by Pauli, then at the Dutch University of Groningen. Years afterwards, Pauli, ever the critic even of himself, wrote to Kronig, "I was so stupid when I was young."[165]

Res Jost had started out at ETH as Pauli's assistant in 1946, the year Pauli returned from the United States. By 1955, he had risen to the rank of associate professor of theoretical physics. He and Pauli were very close. They spoke to each other in the familiar "du."

So it was odd that the four had not yet arrived.

Years later, Armin Thellung recollected the incident. The four men had met for dinner at what Thellung remembers as a "teetotal" restaurant. Afterwards, they set off for Pauli's talk. Here is Thellung:

> Speiser, discovering that the gasoline tank of his Lambretta [scooter] was empty, went to a filling station. There the Lambretta suddenly caught fire! It was extinguished with the water

from a ewer but was not usable any more, so that Speiser had to walk. I found my bike with flat tires and, hence, also had to walk. Kronig, finally, went by tram — a stretch he had traveled many times already — but he forgot to get out at Gloriastrasse, and noticed it only many stops later.[166]

They all managed to find their way just in time, despite the instrumental mishaps.

When the tale was recounted, Pauli was quite amused, perhaps even gratified. The notorious "Pauli effect" — his peculiar and negative influence on all things instrumental — had struck again.

As a theorist, Pauli was on the other side of the divide from experimental physics. Rarely did theorists venture into the laboratory, certainly in the early days of quantum theory. Pauli, though, represented an extreme case, especially among the highly superstitious experimental physicists. It was said that no instrument or laboratory apparatus was safe when Pauli was near. Mechanisms ground to a halt, experimental data disintegrated, glass beakers tumbled to the floor. Thus, "the Pauli effect." The great Otto Stern, who developed the molecular beam for use in studying molecules, forbade Pauli to enter his lab when the two worked in Hamburg. If Pauli came by for a chat or on the way to lunch, he was obliged to knock on the door, to make sure he never set foot within.

By 1940, when Pauli came to Princeton, he had established his name in quantum theory. He remained active, but, as with Einstein, his great discoveries lay in the past. He, too, fit the definition of protégé. At nineteen, at Arnold Sommerfeld's request, he wrote the relativity article for the *Encyclopädie der Mathematischen Wissenschaften,* a project that ultimately turned from an article into a book and garnered praise from Einstein himself. At twenty-two, as an assistant to Max Born, Pauli was invited to the Bohr Festival, his initiation into the vibrant international society of nuclear physics. At twenty-four, he discovered how electrons gather energy levels in the atom — the exclusion principle. Nine years later, he published his paper on what would later be called the neutrino. For his exclusion principle, Pauli was awarded the Nobel Prize in Physics in 1945. The neutrino was another matter. It remained in the realm of theory until 1956, when its existence was confirmed by experiment.

But Pauli's genius, however creative, was circumscribed by his critical acumen. From the mid-1930s to the end of his life, he functioned as the "conscience of physics" — a critic whose often acerbic tongue was widely respected for its impartiality and clarity. He stayed abreast of all the advances and developments; he knew everyone, wrote to everyone, and traveled widely. But after 1933, he made no singular discovery. His contributions to meson theory (the interaction of subatomic particles that "glue" protons and neutrons) and spin (the angular momentum of subatomic particles) were numerous and valuable. Yet Pauli himself seemed unsuited to revolutionary advancement. While he labored over the exclusion principle, his close friend Werner Heisenberg published in 1925 an early formulation of quantum mechanics without fully understanding his own theory. Heisenberg published his "uncertainty principle" in 1927, against the advice of Niels Bohr. Pauli was too cautious and too much of a perfectionist to leap at speculative theories. As collaborators, Pauli and Heisenberg may have combined into the yin and yang of quantum physics. "Pauli's whole character was different from mine," wrote Heisenberg in 1968.

> He was much more critical, and he tried to do two things at once. I, on the other hand, generally thought that this is really too difficult, even for the best physicist. He tried, first of all, to find inspiration in the experiments and to see, in a kind of intuitive way, how things are connected. At the same time, he tried to rationalize his intuitions and to find a rigorous mathematical scheme, so that he really could prove everything he asserted. Now that is, I think, just too much. Therefore Pauli has, through his whole life, published much less than he could

have done if he had abandoned one of these two postulates. Bohr had dared to publish ideas that later turned out to be right, even though he couldn't prove them at the time. Others have done a lot by rational methods and good mathematics. But the two things together, I think, are too much for one man.[167]

A year before his death, Pauli, diffident and critical as ever, quarreled with Heisenberg over a paper on the theory of elemental particles. Heisenberg wanted to publish, despite many doubts; Pauli refused, cautious to the end. The paper remained unpublished.

With his cherubic face and high-spirited demeanor, Pauli stood out among his peers. In a photograph of the 1927 Solvay Conference (a highly selective physics conference held in Brussels), three rows of eminent figures including Einstein, Marie Curie, Bohr, and Max Planck all face dutifully front toward the photographer — all except Pauli, who is looking down and to our left, staring inquisitively elsewhere as the camera clicked.

It was Pauli's first Solvay Conference. At twenty-seven, already famous, he stands in the last row. Front row center was Einstein, then only forty-eight years old, but with white hair, winged collar, and a solemn look. Positioned as they are, one in front and the other at the rear, they embody in that image the course of their working lives. Pauli's entire career coincided with the last half of Einstein's life. Like others of his generation, Pauli moved in the wake of Einstein's epochal discoveries. In spite of their profound disagreement over quantum physics, Pauli and Einstein remained friends. During thirty years of affectionate exchanges, neither wavered, and the great man and the brash youngster never hesitated to voice disagreement.

Certainly, Pauli would have been most welcome in Einstein's living room during that winter of 1943–44. Theirs was something of a transfigured father-son relationship. It was Einstein who nominated Pauli for the 1945 Nobel Prize. At the celebration party, held in Princeton, Einstein declared Pauli his scientific heir apparent.

The usually acerbic Pauli was immeasurably touched. At Einstein's death, he wrote to Max Born:

> Now, that affectionate, fatherly friend is gone. Never will I forget the speech that [Einstein] gave about me and for me in Princeton in 1945, after I had received the Nobel Prize. It was like a king who abdicates and installs me, as a kind of chosen son, a successor.[168]

Still, Pauli was famous for his sharp tongue and combative presence. Even Einstein was not immune to Pauli's verbal assaults. At one lecture, Einstein had just finished making a point when the very young Pauli remarked to the assembly, "What Professor Einstein said is not entirely stupid." One theory, advanced unwisely in Pauli's presence, was so poorly conceived that it was, said Pauli, "Not even wrong!" For the most part, colleagues and students seemed to accept his sledgehammer approach as a reflection of his passion for truth and clarity, at whatever personal cost. Others felt vulnerable. Hans Bethe, for one, gave Pauli a wide, respectful berth after their first meeting at a 1929 conference, when Pauli introduced himself by saying, "Bethe, I was quite disappointed by your thesis." No wonder Pauli earned the nickname "the scourge of God."

Yet he was also a self-effacing colleague and supportive teacher. He fed insights to others, but never thought to claim credit. His letters to colleagues were brilliant small monographs that helped clarify and extend their ideas.

Pauli was born in 1900 in Vienna. But his familial roots — both maternal and paternal — lay in Prague. His maternal grandfather, Friedrich Schütz, was born in Prague, but moved to Vienna, where, as a journalist and playwright, he was active in Jewish political and cultural life. His maternal grandmother, Bertha Schütz-Dillner, was a famed mezzo-soprano, born in Vienna, who sang at the Cologne and Prague opera houses before moving back to Vienna, where she met and married Friedrich. Pauli's own mother, also Bertha, was born in Vienna in 1878. Intelligent and well educated, she was a feminist and a pacifist. She was actively involved in socialist politics. Her writings included theater reviews, essays, and a book on the French Revolution. (Her suicide, in 1927, was one of the precipitating events of Pauli's breakdown in 1931.) Her daughter, Hertha, Pauli's sister, took part in the French Resistance and later became a professional writer.

Pauli's father, Wolf, was descended from Jewish intellectuals who had lived in and around Prague since their expulsion from Spain. At Prague's Carl Ferdinand University, Wolf worked with Ernst Mach, the physicist and philosopher ("Mach" speed was named in his honor), whose works would later directly influence Einstein. After receiving his degree in 1893, Wolf was appointed to the medical staff of Rudolf Hospital in Vienna. There he remained until 1898, when, only months after his father's death in 1897, he changed his name from Pascheles, converted to Catholicism, and married Bertha Schütz, Pauli's mother. He took the name "Pauli" probably in honor of Saint Paul, that earlier convert. The conversion was a prudent and fairly typical step for a professional amid the growing anti-Semitism of Europe. It meant that Wolf, whose specialty was colloid chemistry, would be able to rise to the rank of director of the Biological Experimental Institute and, eventually, full professor and director of the Institute for Medical Colloid Chemistry at the University of Vienna.[169]

Contradictions presided over Pauli's cradle. His father was a Jew who converted to Catholicism to advance his academic career. Young Wolfgang was not a Jew, according to Jewish tradition, since his mother was not (Bertha's mother was Christian, though her father was Jewish). Still, the Nazis would later deem him Jewish (in a 1940 letter to Frank Aydelotte, director of the Institute for Advanced Study, Pauli noted that he was "75 percent Jewish"[170]), and thus his exile to Princeton in 1940.

Even Pauli's baptism was fraught with contradiction. His father Wolf remained friends with Ernst Mach, who had moved to Vienna in 1895. A fierce positivist, dismissing all things metaphysical and spiritual, Mach nevertheless agreed to be the godfather of the new-born Pauli. (Mach seems to have haunted Einstein's living room in 1944. Not only did his positivism lay the foundations for the theory of relativity, it also underlay Russell's view of "neutral monism."[171] Gödel spent his formative years listening to and silently disagreeing with the Machian positivism of the Vienna Circle.) Later, Pauli joked that he grew up a positivist because Mach's power was stronger than that of the baptizing priest. In fact, Pauli was never a simple positivist, though he never ceased to believe that theory must be supported by experiment. His later devotion to Jung and metaphysics would have driven his godfather to distraction.

Like so many other sons and daughter of Vienna, Pauli grew up amid the strange psychic energies, ambivalences, and decadence that created Freud, Ludwig Wittgenstein, Arnold Schönberg, Gustav Klimt, Arthur Schnitzler, and, yes, the young Hitler, who dabbled in art and nursed his monstrous dreams. Perhaps it is not surprising that Pauli later became — at the same time — the disciple of Niels Bohr's crystalline rationality and of Carl Jung's mythifying depth psychology.

Young Pauli attended a "classical" Gymnasium, which emphasized not science but literature and history. He learned Greek and Latin — useful later in his life when he began a lengthy project on Kepler and the alchemists, although his language grades were less than stellar. As for mathematics and physics, Pauli needed little formal training at school to do well. He was a prodigy who had mastered calculus by fourteen and was easily advised in his reading by godfather Mach. His tutor, Hans Adolf Bauer, kept Pauli abreast of the latest theories. At eighteen, just out of high school, Pauli published his first paper, on Einstein's general relativity — a theory published only two years earlier. Whereas many senior physicists were still puzzled by its mathematical difficulties and conceptual innovations, Pauli was unfazed. Even at eighteen, his self-confidence was unshakable. The physicist Victor Weisskopf, Pauli's assistant in the early 1930s, once pointed out to him a mistake in calculation made by another physicist. Pauli said, "Others make mistakes; but I, never." And so it was.

Thus, when he entered the University of Munich, Pauli at eighteen was so far ahead of his peers that his mentor, the eminent theorist Arnold Sommerfeld, put him to work writing the article on the new general relativity theory for the *Encyclopädie Mathematicschen Wissenschaften*. Pauli obliged with a book-length article, 237 pages long, with almost four hundred footnotes (still in print, with supplementary notes added by Pauli just before his death). Upon reading the article, published in 1921, Einstein was all admiration:

> No one studying this mature, grandly conceived work would believe that the author is a man of twenty-one. One wonders what to admire most, the psychological understanding for the development of ideas, the sureness of mathematical deduction, the profound physical insight, the capacity for lucid, systematic presentation, the knowledge of the literature, the complete treatment of the subject matter, or the sureness of critical appraisal.[172]

Such praise from the Master might be the capstone of a career, rather than the starting point. Yet Pauli, now under Sommerfeld's tutelage, blossomed. Munich was, as Sommerfeld wished it to be, a "nursery of theoretical physics."[173] In the early 1920s, it would produce not only Pauli, but Werner Heisenberg and Hans Bethe, each of whom was to win the Nobel Prize in Physics. Still, the city was rocked by political and economic strife as the war ended and the Central Powers disintegrated. Prince Ludwig III, the Bavarian prince regent, fled for his life in 1918 as revolution threatened. In early 1919, Kurt Eisner, a socialist who had been elected premier just a few months earlier, was assassinated. The Communist-inspired Bavarian Soviet Republic lasted only until May, when it was toppled by the Freikorps, many of whom later swelled the ranks of the National Socialists. Munich was, for all practical purposes, the birthplace of Nazism. In 1923, Hitler and his supporters staged the failed Beer Hall Putsch in an attempt to overthrow the fragile Weimar Republic.

Yet Pauli seems to have been oblivious to the turmoil. At nineteen, he was the resident expert on general relativity. He lectured on it in Sommerfeld's class, wrote his second paper on it in June 1919, and tended to sleep late, enjoying the nightlife and clearly untroubled by missing a few morning lectures. He later recalled the "cheerful mood" of those on their way to and from physics and mathematics conferences. Only rarely did the real world intrude on Sommerfeld's institute, as when extremist students threatened to disrupt a lecture by Einstein, scheduled in late 1921. Einstein wisely decided not to attend.

Pauli had already met Einstein at a 1920 conference in Nauheim. It must have been a heady moment for the young student, at work on the relativity article. The conference was one of several during 1920 at which Einstein was expected to defend general relativity. His name recognition not only among physicists, but among the general populace had increased exponentially when in 1919 Arthur Stanley Eddington was able to measure the bending of light during a solar eclipse, thus confirming Einstein's postulate on gravitational magnetism. Still, with fame came controversy, some of it stirred not by science but by blatant anti-Semitism. The Nauheim conference witnessed a "dramatic . . . duel between Einstein and Philipp Lenard," in the words of the mathematician Hermann Weyl. Lenard's anti-Semitism was so virulent that it colored his view of relativity and embittered him against "Jewish physics." Ironically, it was Einstein, the "Jewish fraud" of relativity, who had been able in 1905 to explain anomalies Lenard observed in his own work on cathode rays.

In Sommerfeld's institute, consisting of not much more than a library, a laboratory, a seminar room, the director's own office, and, of course, a lecture hall, Pauli and his fellow students (Werner

Heisenberg among them) learned how to theorize not only on relativity, but on quantum theory as well. It was, of course, the "old" quantum theory first postulated by Bohr in 1913. Bohr's now obsolete atomic model resembled a solar system; its electrons, however, did not follow the rules of classical physics. Sommerfeld was active in the attempt to "manage" Bohr's unwieldy model, suggesting, for instance, elliptical orbits. In the years to follow, Sommerfeld's students, Pauli and Werner Heisenberg among them, rode the wave of quantum theory thoroughly grounded in the rudiments of research. Throughout both of their lives, the very mention of Sommerfeld transformed the usually sardonic Pauli into a deferential and respectful pupil.

In only six semesters, Pauli finished all required coursework. He began his thesis on ionized molecular hydrogen — a little-remembered excursion into quantum theory (Enz remarks that Pauli's ego might have led him to tackle a too-difficult problem). Pauli had distinguished himself sufficiently to be offered an assistantship with Max Born, the physicist whose "probability interpretation" reconciled wave and particle, introduced the notion of probability as a state of knowledge rather than a state of ignorance, and won Born a Nobel Prize in 1954. During the winter of 1921, while completing the thesis, Pauli assisted Born at the University of Göttingen. Until 1933, the university, founded in 1737 by the Hanoverian King George II of England, boasted first-rate mathematics and physics departments: Among other illustrious former faculty was Bernhard Riemann, the nineteenth-century mathematician, whose geometry made Einstein's general relativity theory possible, as we shall see.

Born was fond of Pauli, despite the latter's tendency to sleep late and miss lectures. Much later, Born wrote,

> [E]ver since the time he had been my assistant in Göttingen, I had been aware that he was a genius comparable only to Einstein himself. Indeed, from the point of view of pure science

he was possibly even greater than Einstein even if as an entirely different type of person he never, in my opinion, attained Einstein's greatness.[174]

Genius Pauli may have been; still, his somewhat erratic comportment hinted at a psychological imbalance, which surfaced in the early 1930s.

In Göttingen, Pauli and Born collaborated on an important series of calculations that would, in theory, test Niels Bohr's idea of the harmony of atomic motions. The "Göttingen calculations," based on the celestial mechanics of perturbation (i.e., the effect planets have on each other as opposed to the much greater effect of the sun's gravitation), seemed to contradict Bohr's description of the helium atom. This was one of an increasing number of difficulties facing the Bohr atomic model and his early formulation of quantum theory.

In 1922, Niels Bohr was invited to lecture at Göttingen University. Bohr was at the pinnacle of his career. He had just founded an institute for theoretical physics in Copenhagen, where he taught. He was six months away from being awarded the Nobel Prize for his atomic model. Everyone who was anyone came to the lectures, dubbed the "Bohr Festspiele." Among those attending were Werner Heisenberg and, of course, Wolfgang Pauli. Bohr was to become the greatest mentor of young physicists in the century. Pauli was thereafter a disciple, colleague, and friend of the Danish scientist. The young Heisenberg and Paul Dirac were also drawn into Bohr's orbit.

Bohr's lectures were exciting but not particularly accessible (one student described Bohr's style as "neither acoustically nor otherwise completely understandable"[175]). Still, most of the attendees knew Bohr's theories. The lectures were an occasion to discuss, argue, and augment. Pauli must have been active at the conference, since he wrote Bohr immediately after its close, thanking him for answering "the most diverse questions."[176]

Those questions, together with Pauli's reputation, led Bohr to

invite the twenty-two-year-old prodigy to his new Copenhagen institute for a year. Pauli quickly accepted. In addition to his own
work, Pauli spent the year translating Bohr's papers and lectures
(including Bohr's Nobel Prize lecture) into German. During his
year in Copenhagen, Pauli gained lifelong friends, colleagues, and
collaborators. Above all, he became one-third of the trio who would
forge a new, more successful quantum theory. Though Pauli and
Heisenberg left Copenhagen — Heisenberg went to Leipzig in 1927,
Pauli to Hamburg in 1923 and then to Zurich in 1928 — the three
men met regularly at conferences and corresponded prolifically.

Through it all, Pauli was a "nuclear" force, as it were — not
only an incisive theorist, but a critical sounding board, a mediator,
an adviser. As a collaborator, he supported and inspired, argued
fearlessly, worried the details, and spared no weak postulate his
sarcasm and scorn. Silvan Schweber, reviewing a comprehensive
history of quantum theory, remarks on

> Pauli's staggering contributions to the technical develop-
> ments (Pauli exclusion principle, solution of the hydrogen
> atom in matrix mechanics, spin, paramagnetism, quantum
> electrodynamics, . . .) and to the resolution of the philosoph-
> ical problems engendered by the new mechanics of the micro
> domain. Pauli was the critic par excellence who was at the cen-
> ter of the vast network of correspondents and became the ul-
> timate arbiter of the *Kopenhagener Geist der Quantentheorie.*[177]

Little wonder, then, that Pauli was chosen to write the two
volumes on quantum theory for the *Handbuch der Physik.* The first
volume, published in 1926, summarized old quantum theory —
the state of quantum physics from Bohr's 1913 atomic model up to
1925. The second volume, published in 1933, summarized the new
quantum mechanics and laying out what became known as the
"Copenhagen interpretation." These volumes are bookends to the
heady years during which quantum theory revolutionized physics.

As with all scientific theory, the Copenhagen interpretation was the product of many hands and minds — among them, Erwin Schrödinger, who postulated wave mechanics, Paul Dirac, who devised quantum algebra, and Max Born, who "measured" quantum probability. Still, when the Copenhagen interpretation was explained at the Fifth Solvay Conference of 1927, it was primarily the work of Bohr, the "father" of the new quantum theory, and his two "offspring," Heisenberg and Pauli. Each played his typecast role — Bohr the quixotic and intuitive muse, Heisenberg, the excitable boy wonder, Pauli the indefatigable critic.

It was Heisenberg who devised the linchpins of modern quantum mechanics: matrix mechanics and the uncertainty principle. We will revisit these notions in a subsequent chapter; here, we will simplify dramatically. Matrix mechanics involves measurements of quantum states with a twist: They are not observational measurements. Heisenberg fretted over a simple, undeniable fact: We cannot see into an atom to measure it. If we cannot see the atom, he reasoned, efforts to model it were fruitless (rebuking, as sons do, the father — Bohr was an inveterate visualizer). Instead, Heisenberg set out to quantify the only evidence we can observe: the frequencies and intensities of light spectra. To predict the numerical values of atomic energy, he created a system of equations that, with help from Pauli and Max Born, were extended into a "matrix" language. The pretty atomic model of a nucleus and orbiting electrons had been erased and converted into a numerical table.

While Heisenberg was thinking up matrix mechanics, Pauli was in the grip of the "anomalous Zeeman effect." Named after Pieter Zeeman, a Dutch physicist, the Zeeman effect (as distinguished from its anomalous counterpart) is the splitting of a spectral frequency into three symmetrical lines of very slightly differing energy when placed near a magnetic field. This effect can be explained by classical physics, as it was by Zeeman's teacher,

Hendrik Lorentz. Zeeman and Lorentz shared the 1902 Nobel Prize for their work.

The Zeeman effect held true for some atoms — for instance, hydrogen. In other cases, the splitting resulted not in three symmetrical lines, but in four or more lines that formed complicated patterns: thus, anomalous — and bedeviling. Pauli later recalled:

> A colleague who met me strolling rather aimlessly in the beautiful streets of Copenhagen said to me in a friendly manner, "You look very unhappy"; whereupon I answered fiercely, "How can one look happy when he is thinking about the anomalous Zeeman effect?"[178]

Not until the concept of electron spin would the anomalies be fully explained. Pauli groped on for several years, identifying the valence electron as the culprit in an article first proposing a "classically not describable kind of two-valuedness" of the electron.[179] But what was this "two-valuedness" of the electron? Pauli resisted suggestions of an electron "spin" (thus his deprecation of young Konig) until finally he was convinced, writing to Bohr that he would "capitulate completely."[180]

Bohr's now obsolete 1913 atomic model had, in many ways, done its job: It failed the test of "quantization." Indeed, its visual imprint, based on the solar system, seemed so misleading as to drive physicists away from any proposal that suggested an image. Pauli, a true believer in the quantum, benefited enormously from the swirl of conversation about electrons. In 1924, he read a paper by an English physicist named Edmund Stoner suggesting a distribution of electrons around the nucleus. Pauli was inspired. He set to work formulating the description of electron states that would eventually win him the Nobel Prize.

Known generally as the "exclusion principle," it was dubbed "exclusion" because it describes what *cannot* happen: No two electrons can occupy a single quantum state at the same time. The exclusion principle solved a simple but enormously important question: Why do electrons not fall into the nucleus? The answer is that electrons cannot "fall" into another, less-energized state or "orbital" if it is already occupied. More precisely, no two electrons can have the same quantum states within an atomic structure. These states are expressed as the four quantum numbers: (n) the size or level of the orbit, (l) the orbit's shape, (ml) the orbit's orientation, and (ms) the electron's spin direction. Were electrons not excluded from other states, matter would collapse into itself — as, for instance, in black holes, where the exclusion principle does not hold!

The exclusion principle and electron spin answered a number of questions beyond the anomalous Zeeman effect. Most wonderfully, especially for chemists, the exclusion principle gave the periodic table and its arrangement of elements new meaning. Working empirically, from experimental observations, Dmitry Mendeleyev, the Russian chemist who laid out the table in 1869, had arranged the elements in order of atomic mass. He also grouped them vertically by similar properties. When the exclusion principle clarified the energy value (and position) of the valence shell electron (that is, the outer shell, where bonding takes place), the hidden logic of Mendeleyev's table was revealed. His organization seemed amazingly on target. Those elements with the same valence electrons in a shell are similar — lithium and francium, for instance, each have one valence electron and are both alkali metals, though lithium is light, with an atomic number of 3, whereas francium is quite heavy, with an atomic number of 87.

Pauli's exclusion principle was one of the first in a series of interconnected discoveries that grew out of the "new quantum theory." From 1924 through 1927, Heisenberg, Bohr, Max Born and his assistant Pascual Jordan, Enrico Fermi, Paul Dirac, Erwin Schrödinger, and Pauli, in collaboration or singly, contributed postulates, equations, and theorems to the Copenhagen interpretation. In his role as supercritic, Pauli was Heisenberg's sounding board. Although Heisenberg is credited with creating matrix mechanics, the first complete description of quantum mechanics, Pauli provided the dialectic from which emerged both matrix mechanics and Heisenberg's uncertainty principle of 1927. Indeed, Heisenberg's great insight — that theories must be based on what can be observed — is very Paulian in temperament. While Heisenberg was pondering the "strangely beautiful interior" of atomic phenomena,[181] Pauli obsessed over the anomalous Zeeman effect, determined to explain rather than theorize away experimental observations.

Still, Pauli had time for Heisenberg. Pauli was "generally my severest critic," said Heisenberg. Pauli's feelings toward Heisenberg were more complex. "When I think about his ideas," wrote Pauli in a 1924 letter to Bohr, "then I find them dreadful, and I swear about them internally. For he is very unphilosophical, he does not pay attention to clear elaboration of the fundamental assumptions and their relation with the existing theories. However, when I talk with him he pleases me very much. . . ."[182] Most of Pauli's letters to Heisenberg seem to have vanished during wartime, a particularly sad loss given Pauli's careful, expansive epistolary style.[183] Heisenberg's contribution to the dialogue consisted of thirty-four letters and more than twenty postcards.

Pauli's "staggering contributions" to quantum mechanics continued apace. Yet another oddity reared its head, and Pauli took up the case of the missing momentum. It was a mystery. Experimental data showed that during radioactive processes, the atomic nucleus emits an electron. This is called "beta decay." It occurs because a neutron transforms into a proton. The atom consequently emits an electron. The energy and momentum of all the particles were measured, but the before and after did not match. A tiny amount of energy had gone missing with the beta decay and could not be accounted for. Some physicists, Bohr among them, seemed willing to give up the sacred principle of the conservation of energy. Demonstrably, they argued, energy was lost, and conservation of energy, like much of classical physics, simply did not work for individual subatomic processes.

Not so, said Pauli. He sought the advice of Lise Meitner, a leading authority on nuclear physics. Her work helped him refute Bohr's contention that beta decay did not follow the conservation of energy except statistically — an idea that offended Pauli's austere scientific sensibility. After battling through the possibilities, he came up with a solution — what he called a "desperate remedy." He announced his idea in a letter addressed to the Meeting of the Regional Society in Tübingen: "Dear Radioactive Ladies and Gentlemen," he began. Pauli's wit was famous. The salutation may not have surprised the conferees, but the remedy must have done so:

> [T]here might exist in the nuclei electrically neutral particles, which I shall call neutrons, which have spin 1/2, obey the exclusion principle and moreover differ from light quanta in not traveling with the velocity of light. The mass of the neutrons would have to be of the same order as the electronic mass and in any case not greater than 0.01 proton masses.[184]

Pauli's terminology was soon amended by Enrico Fermi to "neutrino" — in 1932, Sir James Chadwick discovered what he named the neutron, the neutral element equal in mass to the proton. Fermi, unlike the "radioactive" conferees, found the neutrino plausible, since it fit into his theory of weak force and the resulting instability in the atomic nucleus.

Not until 1956, two years before Pauli's death, was the neutrino's existence proven experimentally. It was, said Frederick Reines, its codiscoverer, "the most tiny quality of reality ever imagined by a human being."[185] Today, the neutrino is an invaluable tool in astrophysics. So small in mass and so weak in energy, it passes through the densest material as no other entity can, without collision or effect. Even supernovae, which collapse into unimaginable density, release almost all their energy in the form of neutrinos. Whatever information they carry comes from the very core of the explosion. Pauli celebrated the confirmation of the neutrino with champagne and wrote to Reines and Clyde Cowan in Los Alamos, "Everything comes to him who knows how to wait."

If Pauli's professional work ever suffered from his personal crises, it rarely if ever showed. But crises there were. In 1927, Pauli's father had an affair. His mother's suicide followed soon after. Pauli was stricken, but his anguish remained hidden. The following year, he settled in Zurich, where, despite his reputation as a poor lecturer (his style resembled "a soliloquy . . . often scarcely . . . intelligible" to Markus Fierz, then a student[186]), ETH hired Pauli at the rank of full professor of theoretical physics. In turn, Pauli hired Ralph Kronig to be his assistant for the summer term. They spent the summer as much at play as at work. They were joined by Paul Scherrer, Pauli's nominal department head. Eating, drinking, and concert-going were the usual fare. A favorite haunt was the Kronenhalle, where the famous and infamous among artists (Thomas Mann, James Joyce, Braque, Picasso, Stravinsky) had imbibed.

By the end of 1929, Pauli was married. He had met his bride-to-be, Kate Deppner, at a friend's house in Zurich. She was not, as tradition has it, a cabaret dancer. Rather, she danced at the Max Reinhardt School of Dramatic Arts in Berlin. Pauli probably saw her in Berlin when he visited. In Zurich, Deppner danced at a school run by Trudi Schoop, who later became famous for her work in dance therapy.[187] Schoop later befriended Pauli, heralding his entry into psychology and therapy.

The marriage proved to be a disaster. Kate had been in love with a chemist from Berlin before the marriage, and their relationship did not seem to end with the wedding. As soon as she left Pauli, in 1930, she married her old lover. Pauli's letters during the scant year of their marriage were rueful and self-deprecating. "My wife presumably doesn't join me; even if I am married it is at least in a loose way!" he wrote Oskar Klein.[188] Within two months of the wedding, he promised his friends a printed notice if his wife should run away. Later, according to Enz, Pauli proclaimed himself more distraught at Kate's choice of a "mediocre chemist" as his rival than with the dissolution itself.

With the end of his marriage, Pauli became irritable and subject to mood swings. He drank heavily, spent time in cabarets, picked fights, and was once beaten. During a trip to the United States, in the summer of 1931, he broke his shoulder, having fallen "in a slightly tipsy state,"[189] and had to lecture in Ann Arbor with his left arm uncomfortably elevated in a cast. When depression overcame Pauli in the winter of 1931, his father suggested a visit to the eminent psychologist Carl Gustav Jung. Pauli dived into Jung's works, attended conferences, and made an appointment to see the great man. In February 1932, as Jung proposed, Pauli began therapy with Erna Rosenbaum, a young, inexperienced student whom Jung trusted not to "tamper" with Pauli. Indeed, Jung seems immediately to have recognized in Pauli the makings not only of an interesting patient, but of an inspired collaborator as well. In addition, Jung recommended Pauli to a female analyst — significantly, given Pauli's failed marriage and his mother's suicide.

During Pauli's analysis with Rosenbaum, which lasted about five months, he recounted hundreds of vivid dreams. Some of these dreams he noted in letters to her. In October 1932, Jung took over Pauli's case. For two years, often in letters, Pauli described his dreams to Jung and provided sophisticated analyses of them. "I did not have to explain much of the symbolism to him," said Jung.[190] So rich were the dreams that Jung incorporated many of them into a chapter of *Psychology and Alchemy* (1944).

Pauli and Jung corresponded for decades. Their letters, from 1932 to 1958, are now collected in a book entitled *Atom and Archetype*. Their complex intertwining of ideas and theories is laid out in their joint publication, *The Interpretation of Nature and the Psyche*, published in 1952. Their relationship seems to have grown not so much from Pauli's need for analysis as from their shared intellectual eclecticism.[191] Eventually, the Jungian ideas of synchronicity let Pauli to speculate on a "unified theory" that would not only join the physical forces, but bridge the dualism of physical and psychic that was the hallmark of the modern era. Still, outwardly, Pauli's life ran in separate paths — his colleagues in physics did not know about his immersion in Jungian symbols and archetypes, and his Jungian associates had little contact with ongoing physics.

Analysis seems to have given Pauli a measure of psychological peace and social ease. In 1933, he proposed to Franca Bertram, the woman with whom he would live the rest of his life. "Proposal" might be to grand a word for what Pauli said: "[N]ow we marry."[192] Before their wedding, the couple visited family and friends in Vienna and Hamburg and stayed with Bohr and his family in Copenhagen. Evidently, Franca Pauli enjoyed socializing as much as her husband. Throughout their life together, they entertained and visited friends, many of them Pauli's academic colleagues, throughout the world.

Pauli's professional life in Zurich was happy, but as Hitler's power increased, the Paulis became increasingly vulnerable. Pauli held an Austrian passport. When Austria was annexed to Germany in 1938, Austrian passports were nullified, and Pauli was forced to take a German passport. Given his Jewish heritage, he was subject to Nazi persecution. Switzerland, neutral but geographically vulnerable, began mobilizing its tiny army in 1939. In 1940, however, Germany turned its armies to the west, and Pauli, rather belatedly, began to act. He had applied for Swiss citizenship in 1939. Prudently, in 1940, he also applied for a visa to the United States, on the strength of an invitation to the Institute for Advanced Study. When he was turned down for Swiss citizenship, he arranged for leave for the winter semester from ETH (Eidgenössiche Technische Hochschule). On June 11, with visas to America, Spain, and Portugal in hand, the Paulis tried to board a plane to Barcelona, via Rome. But Italy had just declared war. The plane was cancelled, and the Paulis were forced to wait in Zurich. Finally, on July 31, the couple left by train through France to Barcelona and then Lisbon.

It was a harrowing journey, recapitulated later in the summer by Pauli's sister, Hertha, who wrote about the experience in her fictionalized autobiography, *Break of Time*. Once in Lisbon, the Paulis boarded a ship to New York, where they were met by John von Neumann.

Pauli had visited Princeton once before, during a tour of the United States in 1935–36 (and had met Gödel briefly on the Atlantic crossing). Now, he was to join the Institute for Advanced Study for an indefinite period. In small-town Princeton, during the five lonely war years, he became very close to Einstein. They were wonderfully matched, personally as well as intellectually. The irrepressible Pauli was not intimidated by Einstein's stature. Pauli was sardonic, earthy, tactless, rough-edged — which is to say, a sort of rude version of Einstein. The elder man felt completely at home with his caustic younger colleague, and he gave as good as he got. "You were right after all," he wrote to Pauli in 1931, conceding a point on quantum theory — and added: "you rascal [*Sie Spitzbube*]."[193] In Princeton, they prospered together.

PART 3

THE UNIVERSE

Physics, mathematics, and the universe — these three words form the angles of a tangled and intimate set of relations. Einstein and Pauli the physicists, Gödel and Russell, the mathematicians — each worked within a science that attempts to describe the actual world. That, at least, was the purpose of mathematics at its inception (Euclid's geometry) and the effect of physics in the nineteenth and early twentieth century.

THE LOGIC OF PARADOX

BEFORE WE TURN TO RELATIVITY, quantum mechanics, and the search for a unified theory, we shall take a brief detour into another world altogether — that of mathematical logic. Like physics, the world of mathematics underwent revolutionary changes throughout the nineteenth century and into the twentieth. In so doing, it virtually merged with the doctrines of analytical philosophy and logicism. Few players in the twin worlds of mathematics and logic were more influential than Russell and Gödel.

There is good reason for starting with mathematics. True,

physics began from observations of the visible world. Yet it evolved through mathematics. From the late nineteenth century on, mathematics became an essential tool of the physicist. Though mathematics was never absent from early modern physics — Newton invented calculus, after all — in the twentieth century, mathematics overtook empiricism as the primary method for generating physics. What Newton could observe (albeit through eyes made keen by the imagination) in a falling apple or a setting moon no longer mattered in twentieth-century physics.

Mathematics does not describe the physical world per se. It does, however, problem-solve in the realms of space, number, form, and change. Through mathematics, Einstein explored four-dimensional geometries never seen on land or sea. Today, the mathematics of string theory yields nine space dimensions. These are not observable phenomena. The nine space dimensions cannot even be explained properly in nonmathematical terms.[1] We are led to these proposals not through observation, but through mathematics. Einstein, a born physicist schooled in nineteenth-century empiricism, approached mathematical formalism with trepidation: "As far as the laws of mathematics refer to reality, they are not certain; as far as they are certain, they do not refer to reality."[2]

For pragmatic physicists, mathematical formalism either works or not. Mathematics is a tool. Why does it work? No one has a satisfactory answer. In his celebrated paper, "The Unreasonable Effectiveness of Mathematics in the Natural Sciences," the physicist Eugene Wigner pondered the seeming miracle of the mathematics-physics connection:

> The mathematical formulation of the physicist's often crude experience leads in an uncanny number of cases to an amazingly accurate description of a large class of phenomena. This shows that the mathematical language has more to commend it than being the only language which we can speak; it shows that it is, in a very real sense, the correct language.[3]

Mathematics, like experimentation, sometimes yields surprising or even unwanted results, as if it, too, were beyond human control. In 1928, the British physicist Paul Dirac formulated an equation only to find that it predicted a hitherto unknown and startling particle, the antielectron (or positron). One might even say that it was not Dirac, but his equation (via a minus sign), that discovered antimatter.

Alongside the brief and elegant Dirac equation, general relativity, with its phalanx of equations, is positively epical. It describes not a particle, but the structure of space-time in the universe. It is, nevertheless, a theory tied to observable phenomena, though the observation took place by way of Einstein's "thought experiment" as he imagined himself flying on a beam of light. But, as with Dirac, Einstein's relativity equations were wiser than their maker. As he pondered general relativity, Einstein realized, to his dismay, that the equations described an expanding universe. That was not his intent. To remedy matters, he proposed an emergency fix, a "cosmological term" or constant to keep the universe static. Not only was this fix ill received; it was also wrong. Twelve years later, Edwin Hubble proved that, far from being static, the universe was expanding. The cosmological constant was, in Einstein's view, the "greatest blunder" of his life.[4] It was a blunder born of his preference for the physical and observable over the mathematical. In time, Einstein grew more trusting of mathematical formalism, but only because there seemed no other way to pursue his unified field project.

For Russell and Gödel, no such "practical" matters intruded into mathematics. Still, their work would lead to very practical ends. Out of Russell's system of logical notation and, even more importantly, Gödel's incompleteness theorems emerged the foundations for the computer revolution.

By the time Russell came to Princeton in 1943, he had, by his own admission, left mathematics and mathematical logic far behind.[5] Still, his *Principia Mathematica* expressed the sheer prowess of predicate logic as much through its comprehensiveness as through its innovations. These included an improved notational system and a comprehensive "type" theory based on a hierarchy of "classes." The *Principia Mathematica* inspired successive generations of twentieth-century philosophers: Wittgenstein, Rudolf Carnap, A. J. Ayer, W. V. Quine, and, indeed, the whole of twentieth-century analytical philosophy.

Twenty years later, Gödel, fresh from his dissertation on the completeness of first-order logic, formulated two proofs on second-order logic. (First-order logic differs from second-order in its relative power: First-order logic deals only with individuals or types; second-order logic deals with propositions about individuals or types.) Gödel's two proofs became known as his "incompleteness" theorems. They brought about a paradigm shift as radical as those of "relativity" and "uncertainty." At first, the shift was scarcely noticed. When Gödel made his announcement at a conference on epistemology, only one participant, the brilliant polymath John von Neumann, had an inkling of what the proofs implied. Only very slowly did their depth and breadth sink in. Still, true to the paradigm of paradigm shifts, Gödel's theorems met with resistance from a whole generation of logicians. No wonder: In two proofs, he had demonstrated without any doubt that (1) any consistent formal system that includes the arithmetic of counting numbers (that is, arithmetic using simple cardinal numbers — 1, 2, 3, etc.) is incomplete, as there are statements in the system that cannot be proved or disproved, and (2) a formal system that includes the arithmetic of counting numbers cannot be proved consistent within the system itself.

Gödel's proofs rocked the mathematical world, but it is not useful to exaggerate their effect. They did not cast logic or mathematics onto the garbage heap. On the contrary, logical systems were useful before Gödel's proofs, and they remained useful afterwards. Mathematics continued to depend on axioms and systems that, although "incomplete," worked quite well. But absolute consistency and completeness, much sought as measures of the strength of mathematical systems, could not be found. Like "uncertainty," Gödel's "incompleteness" suggests the limits of what can be formalized. It is always possible, according to Gödel's proofs, to find an axiom that is true but that cannot be proven within the arithmetic system. "Incompleteness" has as its positive formulation "inexhaustibility," argues the Swedish mathematician and computer scientist Torkel Franzen. Gödel himself recognized the philosophical implications, carefully italicizing what cannot be:

> It is *this* theorem [the second incompleteness theorem] which makes the incompletability of mathematics particularly evident. For, *it makes it impossible that someone should set up a certain well-defined system of axioms and rules and consistently make the following assertion about it: All of these axioms and rules I perceive (with mathematical certitude) to be correct, and moreover I believe that they contain all of mathematics.* If somebody makes such a statement he contradicts himself. For if he perceives the axioms under consideration to be correct, he also perceives (with the same certainty) that they are consistent. Hence, he has a mathematical insight not derivable from his axioms.[6]

If a formal system of arithmetic cannot be complete, nor proven consistent within itself (the negative formulation), then it must be (in theory) always open to another axiom, ad infinitum (the positive formulation).[7]

To explain "incompleteness," we must look (briefly) to its place in philosophical history and (more briefly still) at what the proofs achieved.[8] Incompleteness takes its place in — or, more precisely, responds to — a line of philosophical thought that began with Gottfried Leibniz, a true polymath whose expertise ranged from Chinese history to library science. Leibniz was Newton's contemporary and greatest rival: They discovered calculus simultaneously and independently. Leibniz once postulated that space was relative; Newton won that one, and space remained absolute until Einstein. As a logician, though, Leibniz was a towering presence. Into logic, Leibniz injected mathematics. The result was a symbolic logic that would unite mathematics and philosophy.

Important though Leibniz seems to us today, his works fell into some obscurity after his death. He was rediscovered by Kant, the last great Enlightenment philosopher, and then again at the end of the nineteenth century. In 1900, Bertrand Russell published *The Philosophy of Leibniz*. It was, as Russell writes, a "reconstruction of the system which Leibniz *should* have written"[9] had he not been writing philosophy primers for princes. Thus did Russell identify Leibniz as the progenitor manqué of symbolic logic.

Logic became a vibrant and fashionable field towards the end of the nineteenth century: Giuseppi Peano, Ernst Schröder, and Charles Peirce, along with Gottlob Frege, were instrumental in its development. Their project was to rid mathematics of its cobwebs and clutter — ambiguities that in the past had worried no one. Now, in the era of science, mathematics must be systematized and established within the realm of logic. It must be demonstrated, as Russell said, that "all pure mathematics follows from purely logical premises and uses only concepts definable in logical terms."[10] Logicism was born.

Much influenced by Peano, Frege, and nineteenth-century formalism, Russell and his fellow Englishman Alfred North Whitehead launched the ambitious, almost foolhardy project that Frege's system of symbols began. Their plan: to derive all mathematical truths from a set of axioms and inference rules written in symbolic language. The idea grew out of the International Congress of Mathematics of 1900, held in Paris. At that same conference, though not noted by Russell in his autobiography, another germ was planted. It was the celebrated Hilbert "challenge" for young mathematicians: twenty-three mathematical problems that called out for solution. What Hilbert hoped for was a mathematics without paradox.

Paradox there was. In 1902, Gottlob Frege had just published the first volume of his two-volume treatise *Grundgesetz der Arithmetik* (*The Basic Laws of Arithmetic*). In it, he proved that mathematics could be reduced to logic — or so he thought. Russell would prove otherwise by coming up with what was soon dubbed Russell's paradox.

The paradox came to him while he was at work on what would become his monumental *Principia*. Suddenly, he experienced "an intellectual setback."[11] Thinking about sets, he began to wonder about the "sets which are not members of themselves." That would seem a simple concept. The *set* of all red convertibles is not a member of itself. However, the set of sets with more than one member is a member of that very set. What, then, of the set of all sets that are not members of themselves? Is the set of such sets also a member of that set? If so, then it posed a contradiction: If a set of all sets that are not members of themselves is a member of itself, then it is not actually a set of nonmembers, and vice versa. He pondered the contradiction for months, hoping to find a way out. In the end, he broke the news of the paradox to Frege in a letter. Frege's response, though gracious, left no doubt that he felt his life's work had been cast into disarray.

Nothing bedevils mathematicians like paradox. Hilbert attempted desperately to stave off all paradox or, better yet, banish it from mathematical systems. Russell, too, looked in horror upon the paradox he had discovered. His answer was laid out in a work that consumed him and Whitehead for the better part of a decade. *Principia Mathematica* began as a projected one-year, one-volume work. It grew hydra-headed into three volumes (a fourth was planned, but never realized).[12] In it, Russell and Whitehead devised a system of symbolic notation, demonstrated the power of logicism, introduced notions of prepositional function and logical construction, and set the stage for Gödel's metamathematics. To get around his own paradox, Russell proposed a hierarchy of sets: from simple members, to sets, to sets of sets, and so on. No intermingling was permitted in the hierarchy, thus preventing a "set-of-all-sets not-members-of-themselves" logical catastrophe.

Logicism, thus bolstered, spread throughout the philosophical world. In Vienna, especially, there developed a philosophical school of mathematical logic called the Vienna Circle. Its purpose was to found a modern approach to logic that would rid philosophy of metaphysics. Its founding members were Moritz Schlick (who would later be assassinated by a former student on the steps of the University of Vienna), Hans Hahn, Herbert Feigl, and Rudolf Carnap. As the circle grew in number and stature, it moved from smoke-filled Vienna cafés to the university. Meetings were held regularly on Thursday evenings.

In the midst of these antimetaphysicians came a Platonist in sheep's clothing: the young Kurt Gödel, invited as a matter of course by his dissertation adviser, Hans Hahn. He attended regularly for two years, beginning in 1926. As he sat in the shabby classroom observing the birth of logical positivism, Gödel kept his own counsel. His contributions were brilliant, but few. In the end, he let his proofs speak for him. It is ironic, even paradoxical, that the man who undermined the logic of mathematics was schooled by such devout logicists.[13]

Sometime in 1928, Gödel's attendance tapered off. He was at work on his dissertation, which proved the completeness of first-order (i.e., limited) logic. Then, he seems to have turned to what would become the first incompleteness theorem. On October 7, 1930, at a conference in Königsberg, Gödel delivered his seminal paper. It was the third and final day — an inauspicious time at any conference, usually reserved for low-impact papers on obscure topics or for organizational housekeeping. Always a man of few words, Gödel whittled his announcement down to a mere sentence. His "shining hour," notes Rebecca Goldstein, was more like "30 seconds tops." Subdued, uncharismatic, and, at the time, unknown, Gödel was unlikely to have made an impression. No mention of his sentence made it into the conference proceedings. Great men were at the conference: Rudolf Carnap, the father of logical positivism; Friedrich Waismann, Hans Reichenbach, and Hans Hahn, all members of the Vienna Circle; and John von Neumann, a student of David Hilbert, the grand old man of mathematics and head of mathematics at the University of Göttingen.

The Conference on Epistemology of the Exact Sciences, organized by a group of Berlin positivists, was too modest in scope to have drawn the great Hilbert, widely thought to be the greatest mathematician of his time. Hilbert did attend the umbrella conference of the Society of German Scientists and Physicians, held concurrently in Königsberg. Still, on that third day, as Gödel spoke, Hilbert's presence loomed large. Twice in the previous decades, Hilbert had issued formal challenges to fellow mathematicians and logicians. First, in 1900, speaking at the Third International Congress of Philosophy, he had listed twenty-three critical "problems" that remained "unsettled" and exhorted true mathematicians to find the solutions. The most critical of these problems for our purposes was number two: Prove that the axioms in arithmetic are consistent and that, therefore, arithmetic is a formal system without contradiction. Then, at the 1928 International Congress of Mathematicians, held in Bologna, he lectured on "Problems in laying the foundations of mathematics." Now, in addition to the question of "consistency," Hilbert raised another fundamental question: Is it possible, using the axioms of a system, to prove or refute any proposition that is made in the language of that system? Without such proof, Hilbert acknowledged, mathematical logic was without bedrock.

At the heart of Gödel's proof is the Liar's paradox: "This statement is false" — a version of the Cretan Epimenides paradox: "All Cretans are liars." The important thing about the paradox, for Gödel, is its circular, self-referential structure. By definition and design, the paradox references itself. If Epimenides the Cretan is a liar, then the statement "All Cretans are liars" must be false. So Cretans must be truthful — but if so, Epimenides' proposition, that

Cretans are liars, must be true. Likewise for the Liar's paradox: If the proposition "This statement is false" is true, then the statement must be false. In logic, a "well-formed proposition" is either true or false. The paradox does an end run around this either-or structure. The hermetic seal that ensures a logical system is threatened by paradox. Nearly thirty years earlier, Russell had stopped Frege's philosophical program in its tracks with the paradoxical "set of all sets that are not members of themselves" — either it is or it is not; if it is, then it is not, and vice versa. In the context of "true or false," then, the paradox always contradicts logic. But what if, instead of "true or false," we substitute "provable or not provable"? The paradox is then stripped of its "content," as it were, and made analytical. That is what Gödel did in his proofs.

Gödel thus answered Hilbert's "completeness" question with a resounding "No." Arithmetic may be consistent, but it is not possible to prove that consistency using the tools of arithmetic. In a way, Gödel's discovery seems to work well in the world of common sense: It is not possible to see the whole if one is part of the whole. Only from without can we discern the trees as a forest.

Unlike Heisenberg's "uncertainty," word of which had sped around the world of physics, it took several years before "incompleteness" found its audience. This may have been because the proofs were very difficult to understand, even for mathematicians. But resistance, as we have seen, played a part. When Gottlob Frege received Bertrand Russell's bombshell letter regarding the paradox that upended Frege's arithmetic program, the hapless Frege responded within a day: "Your discovery . . . left me thunderstruck," he confessed. Yet he called it "a very remarkable one . . . [that] may perhaps lead to a great advance in logic, undesirable as it may seem at first sight."[14] It was, as Russell later said, a testament to Frege's "dedication to truth. His entire life's work was on the verge of completion . . . and upon finding that his fundamental assumption was in error, he responded with intellectual pleasure clearly submerging any feelings of personal disappointment."[15] By contrast,

Hilbert's initial response to Gödel was anger, according to Paul Bernays.[16] He made no attempt to contact Gödel or respond to the proof. Yet he must have understood how deleterious the impact on his wish to solidify mathematical logic.

When, more than a decade later, Gödel and Russell met (however briefly) and corresponded (however obliquely), it seemed that Russell did not understand the proofs — indeed, in a letter written in 1963, he confessed to their having "puzzled" him.[17] By mischance — or, more accurately, by means of Gödel's nearly pathological perfectionism — we will never know how Russell might have responded to Gödel's critique of *him*. In November 1942, Paul Arthur Schilpp, editor of The Library of Living Philosophers, asked Gödel to contribute to a volume dedicated to Russell's philosophy. Gödel accepted, to Schilpp's (initial) delight, then proceeded to tinker with the initial draft until the end of the following September, by which time Russell had finished his responses to the other essays and could spare "no leisure" for a reply to this latecomer. Gödel's effort elicited no more than a brief note of general praise for Gödel and an off-handed acknowledgement that the "*Principia Mathematica* was completed thirty-three years ago, and obviously, in view of subsequent advanced in the subject, it needs amending in various ways." Gödel's hopes for a "discussion" were dashed.

THE MECHANICAL WORLD

In 1912, Bertrand Russell asked the question: "Is there any knowledge in the world which is so certain that no reasonable man could doubt it?"[18]

This was, of course, the same question Descartes had asked three centuries earlier. To escape from the deadening authority of traditional theology, metaphysics, and morality, Descartes began by doubting everything: the existence of his body, of other people, of the world, of his own sanity. His "methodological skepticism" left him with but one certainty: He was thinking. Hence, "I think, therefore I am." The operations of his mind, distinct from all other matter and perception, were his starting point. From there, Descartes meant to rebuild philosophy afresh by asserting substantive dualism. He split all reality into mind on one side and matter on the other, with no bridge between them (although Descartes provisionally invoked God to fill the gap). During the philosophical wars that followed, Cartesian rationalists and empiricists like John Locke waged their battles against traditional authority with a flaming, righteous sword. Matter was to be explained only by a deterministic physics, without appeal to morality, religion, aesthetics, or other such "mental" qualities ("value-free," as is now said).

In these wars for truth, modern science stood for doubt. Descartes urged us to doubt all — and doubt became the modern mode. Physics is a highly organized method for doubting appearance. Its purpose is to question what it encounters until what remains is what must be. A microscope strips away what we see to reveal the unseen. The seemingly "real" world is replaced by the scientifically "real" one. That scientifically "real" world, so different from that of the spirit, seems to work with regularity, according to invariable rules. Thus was born the underpinnings of the mechanical worldview — the "classical physics" which held sway until the twentieth century.

In 1874, Max Planck, the young son of a theology professor, entered the University of Munich. He had just graduated from the Königliches Maximilians Gymnasium, where he had excelled at mathematics and astronomy, and where he learned the fundamentals of physics, particularly the law of conservation of energy.

It is not unusual for students to be guided by the biases and visions of their advisers. Planck was the exception. In no uncertain terms did his University of Munich physics professor, Phillipp von Jolly, warn him away from the field. All had been discovered, said Jolly. All that remained in physics was the mopping up of a few dusty corners. Planck's stubbornness was later vindicated when he became the default founder of quantum theory and a Nobel laureate.[19]

Jolly was hardly alone. Most physicists agreed — physics was a done deal. After all, Newton's equations had taken care of gravity and motion. As for the remaining forces, in 1864, James Clerk Maxwell presented his famous set of eponymous equations to the Royal Academy. They identified light as an "electromagnetic wave," and they laid down the basic laws for the forces of electricity and magnetism. The primary forces of nature were thus accounted for. What more could physics do?

RELATIVITY OF TIME AND SPACE

Mathematics is curiously intimate with, and revealing of, physical reality. In modern physics, we have discovered the universe's geometric structure and its knotted energies more by way of skeletal equations than by giant telescopes. The more physics uses mathematics, the more physical reality seems to oblige by offering its deepest secrets.

Common sense tells us that physics depends upon empirical data. Galileo's Tower of Pisa, Newton's prism, Young's double-slit box, Foucault's pendulum: Classical physics looked to the physical world for inspiration and confirmation. After Einstein, whose special and general relativity theories were, as he said, "wrested from Nature," all of this would change. Theoretical physics is subject to the "lure of numbers," argues David Lindley. Although physical theories require experimental verification, mathematical structure can make or break a hypothesis. For Lindley, Einstein's "general theory of relativity is the prime example of an idea that convinces by its mathematical structure and power." The theory has been tested, most notably by Sir Arthur Eddington's measurements of light bending around an eclipse on the island of Principe off Equatorial

Guinea. Yet its authority, Lindley believes, comes primarily from its "beautiful theoretical framework."[20]

How did mathematics gain the upper hand in physics? Perhaps it always had an advantage. Plato's *Republic* disdains the empirical: "Geometry is knowledge of the eternally existent."[21] Aristotle fixed his gaze scrupulously at the physical world. From Plato we inherit the suspicion that numbers are magical. Only when Descartes made the connection between "mathematics and the sensory world," as Arthur I. Miller says, did mathematics (and especially geometry) suddenly emerge as a tool for deciphering the physical world. Pythagoras found harmony in the integers in music and in the spheres. Thus it seemed, too, for quantum mechanics. As Miller points out, Max Born called Bohr's analogy of the solar system to the atom a kind of "magic."[22]

Modern science has long been rightly seen as a dissolvent of all certainties — especially physics, which posited uncertainty and wave-particle dualities, split the once-solid atom, and discovered the esoteric geometries of space-time. Yet physics remains a citadel of eternity in its ever-unchanging numbers. As we shall see, physicists have, in their theories of relativity and uncertainty, discovered mathematical formalisms called "universal constants": the speed of light and Planck's constant, for instance. That is some consolation of an "eternal" kind, though ordinary humanity may not be much moved.

The word "relativity" eventually weighed upon Einstein like an albatross of imprecision. It implies opposition to "absolute" — yet Einstein's relativity theories are anchored to the absolute of absolutes: the speed of light. That speed alone remains absolute; time and space are relative to it. In later years, when Einstein fought his rearguard action against the relativism of quantum mechanics, his early discoveries came back to haunt him. "God does not play dice with the universe," he said. But a devil's advocate might say, with Banesh Hoffmann, that Einstein had loosed the demons himself. His quantification of light via Planck is often deemed the inaugural leap into quantum theory. His special theory, with its equivalences of mass and energy, has as its legacy particle physics, the arena for quantum mechanics. Einstein died believing quantum physics to be incomplete in its description.

It is the blessing of youth that its energy is greater than its foresight. Einstein's miracle year produced four extraordinary papers: the first on light quanta, the second on the size of molecules, the third on Brownian motion and the existence of the atom, and the fourth on moving bodies — "special relativity."

In 1896, the discovery of radioactivity inaugurated a search for the nucleus. In 1898, Marie Curie found two radioactive elements, and Ernest Rutherford started sorting out the alpha, beta, and gamma rays from radiation. In 1903, with Frederick Soddy, Rutherford explained radioactive decay, and, in 1911, he finally discovered the atomic nucleus. This set off the next wave of discoveries: Bohr's quantum theory of the atom in 1913; Chadwick's discovery of the neutron in 1932; artificial fission by the Joliot-Curies; and the explanation of nuclear fission by Hahn in 1939.

In 1905, the young Einstein had launched both relativity and quantum physics. (Planck had discovered the quantum phenomenon, but Einstein started quantum physics by applying the concept to light quanta.) Relativity, special and general, were Einstein's single-handed achievement. But very few physicists specialized in that until after Einstein's death. Einstein worked on it by himself.

Quantum physics, however, attracted a crowd and needed them: The implications went in every direction. Einstein himself remained a most important contributor, continuing to publish important work on quantum problems even while laboring away at general relativity. In March 1916, he finally published the complete gravitational theory; in July, he began to publish three papers on quantum theory. Late in life, he told a friend that he had thought a hundred times more about quantum physics than about relativity. As usual, his thinking took a quite individual turn.

Accounts of Einstein's work usually pass quickly over this

longest and, in some ways, most ambitious part of his career. For one thing, this period can be dismissed as evidence of his decline in genius. Indeed, this last effort turned out to be a failure, having added little to the progress of physics. It opened no paths for the future. Recent unifying attempts go in an entirely different direction.

But the question of what happened in Einstein's search for unity may cast light on a neglected side of science. Science is collective and cumulative. Its processes ensure that even its surpassing contributions will ultimately "fail." We rarely see this side of science. Instead, science is presented as a series of dramatic breakthroughs, new pathways, inventions, new frontiers. True, the biologist Robert Hooke and the chemist Robert Boyle were once in the vanguard of discovery; now they have moved back into the fabric of the grand design. Historians of science know that Boyle discovered the relationship between pressure and volume (Boyle's law), but how many working physicists could fairly describe the achievements of the Swedish physicist Svante Arrhenius, whose work on ions predicted the greenhouse effect? Does it matter? Physics is in many ways a self-erasing discipline, concerned only with the latest leading edge of research.

In later life, Einstein was overtaken by history twice. First, by his personal history: In his forties and fifties, his gifts, quickness, and prowess inevitably faded. This happens to everyone. His extraordinary discovery of general relativity may well have made him too confident that he could then master the intricacies of a unified theory. But scientists are overtaken by history in the special way just noted: Sooner or later, the most surpassing achievements will be modified, supplanted, or rebuilt. Newton's gravitational theory eventually became a special case of Einstein's general relativity. If that could happen to Newton, it could happen to Einstein — and indeed, Einstein predicted that it would.

In the world of drama, only Shakespeare can be said to rival Sophocles. Literary works are unique and their truths timeless. But the same cannot be said of science. If Einstein had not modified Newton, someone else would have sooner or later, with whatever variations. Science guarantees that all its members will be challenged and essentially usurped — though it might be more accurate to say superseded, displaced, corrected, or improved. Einstein was challenged in just this way when quantum mechanics emerged in 1925, for its view of the universe contradicted Einstein's most fervent beliefs. If gravity is ever joined successfully to quantum mechanics, even the theory of relativity may well be modified.

Are we to imagine that if Einstein at fifty had retained his youthful powers of imagination, he would have been able to find a unifying theory? That does not seem realistic. No matter how much genius was applied, the time was not ripe: Too little was known about electromagnetism and the fundamental forces of the atom. Strong and weak nuclear forces had yet to be discovered. Yet Einstein, trusting to his formalisms and intuition, dismissed the new evidences of quantum mechanics.

ON THE QUANTUM PATH

In 1905, Einstein, working on a problem called the "photoelectric effect," wrote a paper that some say gave birth to the quantum revolution.[23] This paper, modestly titled "On a Heuristic Viewpoint Concerning the Production and Transformation of Light," offered to solve problems rather than build theory. Still, its language portended radical change:

> It seems to me that the observations associated with blackbody radiation, fluorescence, the production of cathode rays by ultraviolet light, and other related phenomena connected with the emission or transformation of light are more readily understood if one assumes that the energy of light is *discontinuously* distributed in space. In accordance with the assumption to be considered here, the energy of a light ray spreading out from a point source *is not continuously distributed* over an

increasing space but consists of a *finite number of energy quanta* which are *localized at points* in space, which move without dividing, and which can only be produced and absorbed as *complete units*.[24] [emphasis added]

"Discontinuously . . . not continuously distributed . . . quanta" — these are words that fly in the face of Newton and his classical world. It was as if, in his most productive year, Einstein spoke what much later he would, like Shelley's Prometheus, "repent me."

Like all science, quantum physics was built on the shoulders of history. As we have seen, by 1900, the world of physics had split into warring camps or worldviews, each still under the sway of classical Newtonian physics. On one side was the Enlightenment faction, which believed the world a clockwork mechanism. On the other side were the converts to electromagnetic theory, which under Maxwell had wedded electricity and magnetism into a unified theory. Yet problems remained that could not be explained by either side. Among them were three bedeviling conundrums: blackbody radiation, the photoelectric effect, and bright-light spectra, which could neither be explained by classical physics nor ignored (certainly not by young physicists out to make their name). Solving them led inexorably to the quantum revolution.[25]

The first inkling of quantum theory came from a lab in Berlin. In 1900, Max Planck was a forty-year-old physicist with expertise in chemistry and thermodynamics. He was also a fervent believer in the second law of thermodynamics, which states that in a closed system, entropy (loosely translated as "disorder," but also meaning heat loss) increases and, once achieved, cannot be reversed. It was Planck's appreciation of this law and his refusal to give it up that led him to the black-body solution.

Planck was one of the few theoretical physicists amid the cadres of experimental scientists populating German universities.

He was to some extent off the academic radar, and thus had the freedom to contemplate problems that spanned disciplines. Although focused on thermodynamics, he knew of electromagnetism. Maxwell, remember, had demonstrated that light is an electromagnetic wave. Planck believed in Maxwell's findings. More obviously, he noted in the black-body question the intersection of heat (his field), light, and electromagnetism.

Planck set out to examine black-body radiation in the context of the second law. Again, the black-body problem had resisted explanation by classical physics, but held out much practical promise. The reason is that radiation is emitted from the black body in the form of light — specifically, color. For centuries, potters had observed that heat within their kilns turned colors, like a spectrum. From blue through white, each successive color indicates hotter temperature. What can black-body radiation tell us about the behavior of radiation?

As was their wont, German physicists tackled the problem experimentally. They created a black body — an enclosure that would absorb all the electromagnetic radiation it could — with a small hole through which electromagnetic waves could escape. Then they observed the color distribution of radiation coming through the hole. They hoped in this way to study the electromagnetic waves within, just as Maxwell had studied heated gases. To be sure, the question was of more than theoretical interest. Electricity was big business at the turn of the century. If a means for measuring emitted energy could be discovered, electrical companies would be able to quantify their product and provide the greatest amount of power using the least energy.

Two formulas emerged. Unfortunately, one worked well for high frequencies, but not for low frequencies; the other worked only for low frequencies. In fact, at higher frequencies, the second formula produced an impossible result. Light, as had been established, comes in waves, and waves, unlike particles, can multiply infinitely — they just get closer and closer together. As the waves moved closer and closer at higher frequencies, the power (or temperature) would, theoretically, enter the ultraviolet zone and beyond, to infinity. It would become an "ultraviolet catastrophe," emitting radiation with infinite power! Fortunately, nothing like that happens in real life. The black-body heat finds equilibrium, just as Maxwell's gas had. The problem was how to formulate an equation that would explain what was happening.

Planck tried formulae that were tied to the second law of thermodynamics, using standard theories of radiation. Nothing worked. At last, he tried a thought experiment. What if, instead of waves, the black-body chamber was full of oscillating and discrete charges? As the interior heats up, the charges would continue to oscillate at all of the possible frequencies. Planck reworked his formula to fit the experimental results, using a constant to make the equation work. With great consternation, he pondered the result. Only by imagining the electromagnetic waves as discrete elements, using statistics, as Maxwell had with heated gas, and ascribing to the resulting discontinuity a constant (h), could he fit the formula to reality. Planck, forever the enemy of what would become quantum physics, had "quantized" radiation, at least within the black body. At the time, however, he preferred to think of his "constant" as a useful trick rather than a key to atomic architecture.

Ironically, it was Einstein, implacable foe of later quantum theory, who established a theoretical basis for Planck's constant. In essence, he quantized light. He also solved a second conundrum: the photoelectric effect. As Planck was pondering the black-body problem, another German, Philipp Lenard, was experimenting with cathode rays and light beams. He tried shining a light of a single frequency onto a thin metal foil. The result was startling, to say the least. Out of the foil came electrons. The light had somehow ejected electrons. If light were a wave and the electrons were ejecting because they were being disturbed by the energy of the light, then it follows that if the light were of greater intensity, the electrons would carry more energy when they were ejected. But Lenard found otherwise: At low frequencies, even with a very bright beam, no electrons were ejected; at increasingly higher frequencies, the energy of the ejected electrons remained the same. Nothing in classical physics explained Lenard's findings.

The explanation came in Einstein's second paper published in 1905. If, Einstein argued, we consider light not as waves but as photons, we can then explain the photoelectric effect — the emission of electrons that occurs when light is shined on metal. If light acts not as a continuous wave, but as a collection of particles, then the photoelectric effect is nothing more than photons colliding with electrons — tiny particle colliding with tiny particle. His idea of photons explained another problem: that of cooling bodies, which gave off heat not in a neat, continuous way, but discontinuously, "jumping" from temperature to temperature. Newtonian physics could not account for this phenomenon. Quantum physics did. For his discovery of photons — not for relativity — Einstein won the Nobel Prize.

It was not long before someone — it turned out to be the French aristocrat turned experimental physicist Louis de Broglie — asked the obvious question: If light waves behaved like particles, could particles of matter behave like waves?

Before we hear de Broglie's answer, we must take a detour into the atom itself. As Einstein was investigating the large story of gravity and the universe, others searched in the opposite direction, trying to understand the architecture and behavior of invisible particles.

The idea of the atom was first proposed in fifth-century Greece by the philosopher Democritus. If one chisels away at a rock, he reasoned, one is left, eventually, with a fragment so small that it cannot be divided again. These are *atomos* — the Greek word for "indivisible." In the battle over scientific theory, Democritus lost out when Aristotle sided with Empedocles, who defined matter in terms of the four basic elements of fire, water, air, and earth. The atom was lost for more than a millennium. When it resurfaced, in the seventeenth and eighteenth centuries, science found its way back to the atom through the successive findings of Nicolaus Copernicus, Isaac Newton, Christian Huygens, Robert Boyle, Daniel Bernoulli, Joseph Priestley, and Antoine Lavoisier. In 1778, Lavoisier renamed the gas Priestley had isolated "oxygen." It was the first element to be isolated and named.

Then came John Dalton, a teacher and scientist in Manchester, a city at the heart of the English Industrial Revolution. Blessed with typical British weather, Manchester was an ideal location for Dalton, a keen observer who kept meticulous notes, to study fog. He knew from Lavoisier that oxygen combined with hydrogen to make water. In fog he found clarity: Water could behave as air, just as it could ice. What made this possible? The answer was — atoms. In air, the atoms were spaced far apart; in solids, atoms bunched together. For the next century, scientists discovered, analyzed, and classified elements. Still, the atomic structure, by definition invisible, remained a mystery.

Toward the end of the century, the veil began to lift. At Cambridge, a young mathematician, J. J. Thomson, was put in charge of the Cavendish Laboratory. Under Thompson, the Cavendish flourished, attracting first-rate students and researchers. Its fame was solidified in 1897, when Thompson discovered the electron (which he called "corpuscle") by isolating the particles that make up cathode rays. With the venerable Lord Kelvin, Thompson proposed a rather chunky atomic structure, a souplike concoction with floating electrons, dubbed the "plum pudding model."

As one might expect, the plum pudding model found few backers besides Thompson and Kelvin. Fortunately, the Cavendish nurtured great students. In 1895, when applicants from abroad were first admitted, Ernest Rutherford, fresh off the boat from New Zealand, appeared at the door.[26] The experience of working with Thompson changed Rutherford's life. He became an atomic specialist, landed at the University of Manchester, and, in 1909, conducted a "most incredible" experiment. With his students Hans Geiger and Ernest Marsden, he shot alpha particles (bundles of neutrons and protons emitted by radium) through a thin sheet of gold foil. Most of the particles passed through. A few, though, bounced back. The plum pudding model had no hard centers to stop the alpha particles. How to model this phenomenon? Rutherford borrowed the image of the solar system, with electrons circling an interior nucleus. The Rutherford model was not without problems. Still, it "worked," just as Newton's gravity had. By envisioning the solar system model, Rutherford and his students measured the nuclei of different elements. They could now explain atomic number and nuclear weight with much greater clarity. Over the next few years, Rutherford looked deep into the atom, and in 1917 he became the first scientist to "split" the atom by bombarding a nitrogen nucleus, transforming it into oxygen and emitting hydrogen. The "solar system" model is still with us. It is useful and easy to visualize.

As Rutherford had revised Thompson's plum pudding model, so would a Rutherford student rethink the atom as solar system. Niels Bohr came to Manchester in 1911, armed with a complete set of Dickens from which to learn English. He had a doctorate from Copenhagen University and an impressive background in electron theory. Little wonder that he had sought out Rutherford's laboratory. Rutherford was an ideal teacher — cheerful, avuncular, and inspirational. His laboratory, if a bit rollicking, teemed with ideas and energy. He was known to sing "Onward Christian Soldiers" to his student-troops, his booming voice preceding him as he swept from room to room.

At Manchester, Bohr tackled the inherent problem of the solar system model with typical Continental audacity. He knew that Rutherford's model was wrong according to classical physics. An electron circling the nucleus would emit energy (because of angular momentum) and thus fall into the nucleus. The atom would collapse, and matter would not exist. Bohr stabilized the model by abandoning classical physics. His electrons would move in fixed orbits around the nucleus. Each orbit corresponded to an energy level. The lowest energy level was closest to the nucleus.

To reach these conclusions, Bohr himself made a quantum leap. If, rather than continuously emitting energy, the energy loss, like Planck's quanta, is discrete and particle-like, not continuous and wavelike, then electrons would emit fixed amounts of radiation when they move from one orbit to another. This "jump," Bohr reasoned, is as discontinuous as Planck's black-body charges and Einstein's photons. The momentum of a particle changes (rises or falls) in discrete quantities. In other words, like Isaac Asimov's spaceships, electrons "jump" instantaneously through space, from one orbit to another. When an electron jumps from a higher energy orbit to a lower one, it emits light. When an electron jumps from a lower energy orbit to a higher one, farther from the nucleus, it does so because it has absorbed energy from some other source. This happens, for instance, when a chlorophyll molecule in a maple leaf or the metal hood of a black SUV absorbs light. The chlorophyll molecule absorbs heat and converts it into food for the tree; the black SUV atoms radiate heat, electron by electron, sufficient to fry an egg. They do so not by emitting heat continuously, but discontinuously, by emitting "quantum" amounts of heat generated when an electron "jumps" from a higher to a lower energy state.

Discontinuity is the key concept here. No longer was physics solely within the classical realm. The quantum moment had arrived. Into the fray stepped a new generation of young theorists unattached to classical physics and chafing at its inadequacies. Of these, Pauli, Heisenberg, Paul Dirac, Louis de Broglie, and Max Born stood out. In rapid succession, from 1914 to 1927, came the building blocks of quantum physics: confirmation of stationary solid states (James Franck and Gustav Hertz); confirmation that matter was both particle and wave (Arthur Compton and de

Broglie); Pauli's exclusion principle; matrix mechanics; and two sets of statistics for counting particles (Bose-Einstein and Fermi-Dirac).

Far from settling matters, though, these discoveries demolished Bohr's atomic model. Its death knell sounded in 1924, when de Broglie's doctoral thesis proved that matter was not just particles, but also waves. He did so in part by applying to all matter the lessons of Einstein's photon. The "pilot" waves that follow matter through space are not incidental, but have frequencies directly related to the particle's motion.

Matter, like radiation and light, now possessed this dual nature. No longer was physics divided into two camps, as de Broglie remarked in his Nobel Prize speech. The two conceptions of physics — matter, governed by Newtonian mechanics, and radiation, envisioned as traveling waves — were now "united."[27] Bohr's model, shackled to the image of electrons as particles, no longer stood. Its demise plunged the subatomic world into the same state of disarray that had befallen macrophysics with Einstein's theories of relativity. Suddenly, our intuitive sense of the world no longer held true. Beneath (and above) our world of appearances, there exist wholly different worlds. In one, all motion is relative except for the speed of light. In the other, particles are waves and the reverse, obeying laws that contradict even Einstein's revolutionary laws.

THE COPENHAGEN INTERPRETATION

Into the breach of Bohr's atomic theory, now in tatters, stepped Werner Heisenberg. Fresh from a year of apprenticeship with Max Born, Heisenberg was acknowledged to be a brilliant theorist with an aversion to experimental physics.[28] He and Pauli met Bohr at the Göttingen lectures of 1922. So taken was Bohr with Heisenberg's questions that he proposed a walk up Hain Mountain. During that afternoon, wrote Heisenberg, "my real scientific career . . .

began."[29] It was the first of many conversations, often heated, between the father of quantum physics and the daring and inspired Heisenberg. Pauli often served as intercessor when disputes between the two threatened progress. It worked. Together, Heisenberg and Bohr forged a complete theory that would become known as the "Copenhagen interpretation."

Still, it was a tangled relationship — one that became more tangled in 1941 when Heisenberg and Bohr took their famous evening stroll through a park in German-occupied Copenhagen. At that meeting between the German patriot and the Danish Jew, Heisenberg did or did not try to extract from Bohr atomic secrets; did or did not hope to discover the extent of the Allies atomic program; did or did not suggest the immorality of atomic weapons. What happened during that evening stroll has been a matter of dispute ever since and fodder for Michael Frayn's play "Copenhagen." (That their meeting took place in a woodland is in itself interesting. Nature was the backdrop for several "leaps" in quantum theory, most famously when Heisenberg's hay fever forced him into seclusion on a North Sea island, where he pondered atomic structure and thought up matrix mechanics, the first formulation of quantum mechanics.)

In the happy years of the 1920s, Bohr played the diffident, sometimes disapproving father, Heisenberg the rebellious and brilliant prodigy. In his three years at Copenhagen, from 1924 through 1927, Heisenberg proved his worth. His first major contribution was a formula that figured the energy states within an atom. Max Born and Pascual Jordan extended the formula into a true matrix mechanics with which all frequencies of the spectrum could be figured. Heisenberg's second contribution was less formalistic and much more incendiary. The uncertainty principle by its very nature contradicted classical physics and, in a way, challenged the very essence of modern science. Throughout the development of classical physics, it was assumed that perfectly accurate measurement was possible, in ideal conditions. Only the crudeness of our measuring apparatus stood between our results and the object's true dimensions. Heisenberg said no, it is impossible to "see" sufficiently into the atom to know for certain what processes — specifically, wave or light — one is measuring. Further, whatever means we use to measure will inevitably disturb the element under scrutiny. Thus, the "observer" will affect and distort the "observed."

For Bohr, Heisenberg's uncertainty principle was too limited. It focused only on conditions that occurred during observation, and in effect ignored the "wave" state. Bohr insisted (vehemently) on much more. Little was gained, he argued, from ignoring empirical evidence.

The wave-particle duality that so vexed quantum physics had a very solid empirical history. It began with an experiment, now replicated in physics classrooms throughout the world, called the "double-slit experiment."

In 1801, Thomas Young (a physicist, physician, and Egyptologist who in his spare time deciphered the Rosetta Stone) devised a mechanism to analyze light: "I made a small hole in a window shutter, and covered it with a thick piece of paper, which I perforated with a fine needle," he told the Royal Society in 1803. Thus began what has been called the most beautiful experiment in physics.[30] When Young "split" the sunlight by dividing it with a thin card, he observed, projected onto the wall, "fringes of colour on either side of the shadow" — clear evidence of interference or diffraction. He was astonished. Light, according to Newton, was made of particles. Yet as it traveled past Young's thin card, it diffracted, just like a wave breaking on a jetty. Startling as this was for Young, a 1927 variation on the double-slit experiment came up with an even more astounding result: Electron beams from a nickel crystal produced diffraction. Matter, like light, was shown to behave like waves.

Young's experiment was the basis for much resistance to Einstein's theory of photons. After all, Young had proved that light was wave, not particle. Einstein countered with proof of light's particle nature. In 1915, Robert Millikan, after ten years of experimentation, reluctantly concluded that Einstein's equations were correct. In 1923, Arthur Compton confirmed the particle nature of electromagnetic quanta by observing the scattering of electrons from X-rays.

Thus, the paradox: Light and, as de Broglie proved, matter are both wave and particle. If physics had challenged our intuitive sense of the world with relativity, it now seemed to have done away with intuition altogether. After all, these states — wave and particle — are quite different. A wave is continuous and nonlocal, spread out over a large area. Particles are discrete, indivisible, and local. Richard Feynman calls the uncertainty principle a "logical tightrope on which we must walk if we wish to describe nature successfully."[31] We must think anew, says Feynman, "in a certain special way" to avoid "inconsistencies." The awkward phrasing is unusually revealing. Quantum physics challenges not only our intuitive perception, but the limits of language. When Bohr and Heisenberg argued over the uncertainty principle, more was at stake than the utility of the principle itself. In order for a theory to "work," it must not only explain evidence; it must also gain acceptance among practitioners. Bohr was older than Heisenberg and certainly more empathetic to the general state of alarm over quantum mechanics. Perhaps, too, he was more understanding of our psychological need to visualize (that is, imagine, in its etymological sense) our world. Something in addition to de Broglie's elegant proofs was needed to bridge the gap between particle/discontinuity (the side favored by Heisenberg) on the one side and wave/continuity on the other.

The answer, Bohr came to believe, was "complementarity." Rather than see particle and wave in opposition, as mutually exclusive, might we not accept both as true? Particle and wave are interdependent; so, too, are classical physics and quantum theory. Here, seemingly, was an attempt to forge connections. Yet beneath the gentle-sounding word, complementarity posed a radical idea. Not only was exact measurement impossible, but the ambiguity lay in the properties themselves. Even before measurement, the atomic system is uncertain. All we can hope to attain by way of information lies in the realm of probability.

We have seen that quantum physics developed not through the genius of a single thinker, but through a series of conversations among both believers and nonbelievers. The most intense of these took place between Einstein and Bohr. It began well before Bohr's landmark presentation of the Copenhagen interpretation in 1927. Indeed, Einstein had contributed to the theory himself, not only by "discovering" the quanta, but also in collaborating with S. N. Bose to develop Bose-Einstein statistics (a set of formulae defining the statistical distribution of particles called bosons).

Einstein, unwilling to set foot in Fascist Italy, did not attend Bohr's lecture in Como. Within a few weeks, however, Bohr was scheduled to speak at the 1927 Solvay Conference in Brussels. Einstein came to the conference full of misgivings. What he heard was Bohr's paper, "The Quantum Postulate and the Recent Development of Atomic Theory." He was horrified. In Bohr's words, "Einstein . . . expressed a deep concern over the extent to which causal account in space and time was abandoned in quantum mechanics."[32] Thus began the famous dialogue between Bohr and Einstein, one that would help Bohr refine his theory of complementarity and spur Einstein ever forward in his search for a theory that would subsume quantum mechanics and encompass both atomic physics and astrophysics.

EINSTEIN AND UNIFIED THEORY: CHASING THE RAINBOW

By 1927, Einstein had begun his assault on quantum mechanics. He never let up. It led to the most famous dispute of twentieth-century physics. Two eminent physicists, Albert Einstein and Niels Bohr, remained locked in battle for years, raising questions that remain unsettled even today.

Einstein's stubborn criticism struck many, particularly of the younger generation of Pauli and Heisenberg, as folly, waste, even provocation. In their view, the greatest physicist and boldest pioneer of the age had become a reactionary. The young Robert Oppenheimer visited Princeton in 1935 and described Einstein as "completely cuckoo."[33]

Yet the old Einstein was only acting like the young Einstein. In 1905, he had opposed the orthodoxy of those who thought that light flowed through the invisible material called "aether." Now, he was a "heretic," cast out by the new quantum orthodoxy. He became a prophet in the wilderness, preaching the need to refound all of physics on a revivified "classical" basis, closer to Newton than to Heisenberg. This time, though, he would not succeed.

He failed in part because subatomic physics was in its infancy, unable to yield sufficient data. Physics had to wait years before "powerful atom smashers would clarify the nature of subatomic matter," notes Michio Kaku in his reverential *Einstein's Cosmos*.[34] Had Einstein had our wealth of data, Kaku hazards, he might have succeeded. As it was, knowing nothing of the bosons, gluons, muon neutrinos, partons, leptons, and quarks — the whole atomic stew, as it were — he could not form a "picture" to guide his mathematics.[35]

Yet Einstein's grand project had an agenda, one that skewed his method. He not only disagreed with the new quantum theory; he detested it. Not only did quantum mechanics turn the universe into a game of dice by replacing causality with probability, but it also seemed to him inelegant and ungainly. He was especially repelled by Heisenberg's uncertainty principle and its implied threshold beyond which we must remain ignorant. In his effort to unify electromagnetism and gravity, Einstein remained within the fold of classical physics. However revolutionary his own notion of the relativity of space and time had been, his unified field theory would succeed not by accepting the newest revolution, but by subsuming it.

With anyone less sane and generous than Einstein — and Bohr — the dispute over quantum mechanics could have turned acrimonious. The issues were not technical, but philosophic. Einstein on the one hand, and Bohr, Pauli, and Einstein's close friend Max Born on the other, were arguing from their deepest convictions about what physics should be. When Paul Ehrenfest had to choose Bohr over Einstein, he began to sob. Yet no real personal bitterness or resentment surfaced in this dispute, which, while intense, was respectful, though it lasted for over thirty years. Unable to change each other's mind, Einstein and Bohr finally talked past each other. By the late 1930s, they were exhausted. Afterwards, when they met, they exchanged pleasantries, but no more. Einstein and Born continued to argue for decades. In 1926, Born wrote about the new quantum mechanics to Einstein, and got this reply:

> Quantum mechanics is certainly imposing. But an inner voice tells me that it is not yet the real thing. The theory says a lot, but does not really bring us closer to the secret of the "old one."[36]

Einstein's confident appeal to an "inner voice" upset the rigorous Born, but their friendship never wavered. In 1944, Einstein wrote Born that they had become "antipodean" in their scientific beliefs: "You believe in a God who plays dice, and I in complete law and order in a world which objectively exists."[37] Born answered a month later, suggesting that Einstein was not really up to date with the arguments. "[G]ive Pauli a cue," wrote Born, "and he will trot them out." On one of Born's later manuscripts on the topic, Einstein wrote in the margins such remarks as "Blush, Born, blush!" and "Ugh." To which the astonished Born replied: "Do you really believe that the whole of quantum mechanics is just a phantom?"[38]

In their letters, mutual affection mingles with frustration, especially from Born. In 1948, Einstein and Born exchanged what appear to be plaintive cries in the night:

> (Einstein): I am sending you a short essay which, at Pauli's suggestion, I have sent to Switzerland to be printed. I beg you please to overcome your aversion long enough in this instance

to read this brief piece as if you had not yet formed any opinion of your own, but had only just arrived as a visitor from Mars. . . . I am . . . inclined to believe that the description of quantum mechanics . . . has to be regarded as an incomplete and indirect description of reality, to be replaced at some later date by a more complete and direct one. . . .
(Born): . . . . Your example is too abstract for me and insufficiently precise to be useful as a beginning. . . . I am, of course, of a completely different opinion from you. For progress in physics has always moved from the intuitive towards the abstract. . . . Quantum mechanics and the quantum field theory both fail in important respsects. But it seems to me that all the signs indicate that one has to be prepared for things which we older people will not like.[39]

They kept sparring, and in March 1954, Pauli finally intervened. He was visiting in Princeton again, and Einstein had given Pauli an article Born had written. Pauli, as usual, was blunt in his letter to Born:

[Einstein] is *not at all* annoyed with you, but only said you were a person who will not listen. This agrees with the impression I have formed insofar as I was unable to recognize Einstein whenever you talked about him in either your letter or your manuscript. It seemed to me as if you had erected some dummy Einstein for yourself, which you then knocked down with great pomp.[40]

Pauli's explanation of Einstein's "classical" view was "simple and striking," Born remarked, but as usual, no mind was changed. As quantum theory developed from 1925 until his death in 1955, Einstein went his own way, ignoring the new findings and searching for a unified theory according to his original plan.

Only in Newton do we find a parallel for the first half of Einstein's life: his discovery at age twenty-six of relativity, the identity of mass and energy, and the quantum nature of light, and at age thirty-six of gravitational theory. But no one, not even Newton, rivals the later Einstein. Newton spent the last half of his life pondering theology and running the Royal Mint. Einstein was in his mid-forties when he began his search for the unified field. At that age, Bohr, Planck, and Rutherford, their great discoveries behind them, were slipping away from science into administration or teaching. To the day he died, Einstein toiled away at a problem more ambitious than any he had faced before, as if he still had his greatest discovery in front of him.

Today, Bohr's Copenhagen interpretation, as is the common fate of orthodoxies, has undergone dramatic revision. In a sense, quantum mechanics has been reworked into a near-unified theory called the "standard model," a quantum field theory that is consistent for special relativity and quantum mechanics — that is, consistent for every point in space. Although early proponents of quantum mechanics tried their hand at field theories, nothing succeeded fully until Richard Feynman and others successfully manipulated the necessary mathematics. The result was quantum electrodynamics, or QED, which described the electromagnetic field and the interactions of particles precisely to ten decimal places.

When Einstein started his quest, he meant to unify gravitation and electromagnetism. Since then, the additional forces, strong and weak, have been unified with electromagnetism under QED. Gravity is the odd force out. It has yet to be folded into the other three forces. Any "grand unified theory" (dubbed GUT) or "theory of everything" must connect all four forces: gravitation, electromagnetism, and the two forces, strong and weak, that stabilize the nucleus. These forces have rightly been called primordial. They have existed from the beginning, or as near to it as can be imagined. When matter came to be, it found those primordial forces waiting to turn elemental matter into the universe we know. We do not know why gravity or electromagnetism or the strong or weak forces exist; like the universe, they simply do.

The holy grail of unified theory continues to attract believers. String theories, attempts to tweak the standard model into a form compatible with gravity, have long been contenders. Some hope that an ultimate "superstring" theory will be a "theory of everything." Not yet. Among other problems, no variation of string theory can be verified experimentally. Michio Kaku dismisses the need for experiments:

> The real problem is purely theoretical: if we are smart enough to completely solve the theory, we should find all of its solutions. . . . So far, no one on Earth is smart enough.[41]

In a way, modern physics has always been driven by the ambition to unify. In the view of one recent account, "modern physics is best summed up a series of unifications."[42] In the late Middle Ages, it was assumed that matter on the earth and moon differed from that of the celestial bodies, that each domain was governed by different laws, that each element of the "universe," from the earth to the farthest quasar, was sui generis. Modern physics was born with Copernicus, Kepler, Galileo, and Newton. Each searched for a single system that would apply to all space, time, and matter.

For a time, the solution was at hand. Newton's mechanical system, complete and quite successful, prevailed for more than two centuries. Then, in 1819, Hans Christian Oersted, a Danish professor, noticed the deflection of a compass needle towards a source of electricity. Magnetism and electricity were, somehow, one and the same. Everything changed. Electromagnetism, theorized in its classical form by James Clerk Maxwell, involved phenomena quite different from Newton's solid bodies. Maxwell even demonstrated that light is an electromagnetic wave. Newton's mechanical model began to crumble around the edges. The struggle to reconcile electromagnetism with the forces of gravity, space, and time gave rise to relativity on the one hand and new atomic theories on the other.

Pass a magnet over a heap of iron filings, and the filings line up in a certain direction, as if an enveloping force has moved them all — which is exactly what has happened. The field is the magnet's sphere of influence. Gravity is another such field; its power extends to all objects in its region, pressing us to the earth and preventing us (and everything in our world) from levitating. The clustering of the filings around the magnet's poles illustrates the anatomy of a field. The lines of force tell the filings where to move and how quickly. The field itself generates the power that moves the filings. In Newton, particular forces "told" each body how to move and at what speed and strength — like billiard balls struck by a cue. Maxwell's field, however, simplifies things greatly: It is both particle and force at once — the vehicle and the source of energy.

Einstein sided with Maxwell and against Newton. Indeed, relativity was born from Faraday and Maxwell's "new type of law." In Newton, one body was assumed to affect another instantaneously, no matter how vast the distance. But how could a force be transmitted "instantly" when nothing can go faster than the speed of light? And besides, what "medium" carried forces from the sun to the earth or from the earth to the moon? Until Einstein, physicists postulated an ether through which electromagnetic waves and light traversed. From Maxwell, Einstein began to see that there were a myriad of "local" fields everywhere in space, transmitting energy — a crucial step in the journey toward general relativity.

Einstein's quest for a unified theory was driven by his belief in the harmony of nature: Gravity and the atomic dimension *had* to be the same in some deep underlying way. That same conviction had sustained Einstein during his ten-year search for general relativity. No less a physicist than Max Planck told Einstein that his attempt to generalize relativity was futile. As Einstein biographer Abraham Pais remarks:

> There is no evidence that anyone shared Einstein's views concerning the limitations imposed by gravitation on special relativity, nor that anyone was ready to follow his program for a tensor theory of gravitation. Only Lorentz had given him some encouragement.[43]

Despite the doubts of others, Einstein constructed his theory of general relativity. The self-confidence it gave him can hardly be overestimated. He expected to prevail again in his quest for a unified theory. In 1949, though stymied for a quarter century, he still evoked the heady experience that culminated in general relativity. After cataloguing the "impossibles" evoked by the quantum partisans — that is, the numerous ways in which Einstein's search for unity could not reconcile with the evidences of quantum mechanics — Einstein stubbornly returned to his thesis:

> All these remarks [i.e., against a unified field theory built on relativity] seem to me to be quite impressive. However, the question which is really determinative appears to me to be as follows: What can be attempted with some hope of success in view of the present situation of physical theory? At this point it is the experiences with the theory of gravitation which determine my expectations. These equations give, from my point of view, more warrant for the expectation to assert something *precise* than all the other equations of physics.[44]

Yet it was a false hope, or so it now seems. Einstein wanted to ground his unified theory upon general relativity — that is, his explanation of the force of gravity. Yet in recent history, it has become clear that the ground floor must be quantum physics, not gravity.

As he kept trying, he reversed a bedrock assumption that governed the first half of his career. Before he took up unified theory, Einstein was an empiricist who trusted only his intuition about physical reality, not the beauty or inner consistency of equations. His 1905 paper on special relativity contained very little mathematics — and simple mathematics, at that. Indeed, when a small academic industry soon began to formalize and refine the mathematics of special relativity, the young Einstein mocked the efforts as "superfluous learnedness."[45] In 1918, his mathematical friend Besso apparently suggested that, in the discovery of general relativity, mathematics had been more important than empiric knowledge. Einstein was irked and flatly disagreed:

> You allude to the development of relativity theory. But I think that this development teaches us nearly the opposite: if a theory is to inspire confidence, it must be founded on *facts* susceptible of being generalized. . . . Never has a useful and fertile theory been found by purely speculative means.[46]

In 1918, the mathematician Hermann Weyl made a try at unifying gravity and electromagnetism, and amazingly found apparently successful equations. Einstein crushingly replied: "Your argument has a wonderful homogeneity. Except for not agreeing with reality, it is certainly a magnificent achievement of pure thought."[47] Again and again, Einstein championed "reality," "facts," and "experience."

But as his search for a unified theory dragged on, his views changed. He began to speak of the "pure thought" of mathematical formalism as a uniquely privileged approach to the reality of physics. Here is the former empiricist preaching the new message in a 1933 lecture at Oxford:

> I am convinced that we can discover by means of purely mathematical constructions the concepts and laws connecting them with each other, which furnish the key to the understanding of natural phenomena. . . . Experience remains, of course, the sole criterion of the physical utility of a mathematical construction. But the creative principle resides in mathematics. In a certain sense, therefore, I hold it true that pure thought can grasp reality, as the ancients dreamed.[48]

The phrase "purely mathematical constructions" comes as a shock. In 1921, still skeptical, he had put the matter with his customary lucidity: "[A]s far as the propositions of mathematics refer to reality, they are not certain; and as far as they are certain, they do not refer to reality."[49] Now, twelve years later, he had come to saying the opposite: The more certain the equations, the more they refer to reality. Reversing the view he had expressed to Besso in 1918, he claimed that general relativity did spring mainly from its formalisms: "Coming as I did from skeptical empiricism of [Mach's] type . . . the gravity problem turned me into a believing rationalist, i.e., into a person who seeks the only reliable source of truth in mathematical simplicity."[50]

Einstein had not turned into a mathematician. He had changed his method, not his goal. Mathematics for him was never the point, only a tool for doing physics. When the French mathematician Elie Cartan suggested a promising but complex theory, Einstein replied with his inimitable humor:

> For the moment, the theory seems to me to be like a starved ape who, after a long search, has found an amazing coconut, but cannot open it; so he doesn't know whether there is anything inside.[51]

Still, the popular cartoon image of Einstein as a wild-eyed scientist standing before a blackboard crammed with incomprehensible equations springs from his work on general relativity. Such an image was not possible before Einstein. General relativity required such new and rarefied mathematics that in 1914, the great German mathematician David Hilbert said half-jokingly that "physics has become too difficult for the physicists." After general relativity, theoretical physics and abstruse mathematics were ever more closely wedded.

In 1919, when proof came that the sun's field curved light, Einstein entered the pantheon alongside Newton. Satisfying enough for anyone else, but not for Einstein. He was unhappy that general relativity had produced another dualism at the center of physics — this time, matter versus field. Einstein found such incoherence in the structure of physics "intolerable."

> Could we not reject the concept of matter and build a pure field physics? What impresses our senses as matter is really a

great concentration of energy into a comparatively small space. We could regard matter as the regions in space where the field is extremely strong.[52]

He thus sought an even more heroic theory, one able to dissolve matter into pure field laws. To do this, he had to enlarge general relativity, just as he had generalized special relativity.

Einstein's effort to find a unified "theory of everything" so clearly mirrors his earlier work that a brief look back at a few points is indispensable:

The famous perplexities of special relativity (1905) — peculiar clocks, shrinking distances — arise from a situation we do not encounter in our commonsense world. If we drive our car from New York, we know how many miles we are from that city, where it is, and how long it will take to drive back. We move about; New York is fixed in place.

But what if New York were also moving, as well as all the landmarks in between? How would we know exactly where we are and when something happens? Or, put otherwise: How can we do the physics? Special relativity accepts that all matter is in constant motion: Galaxies, planets, and all observers thereon are moving relative to what they observe. In this universal flying circus, there is no privileged space from which to measure, and no "absolute" time from which to count. Since no one can freeze all motion to get things utterly straight, each observer inevitably sees "simultaneous" events differently. Einstein's genius was to understand how we can nonetheless get an accurate measurement of time and distance, without which physics is stymied. First, he proposed that differences can be aligned to ensure that the same laws of physics take the same form wherever observed.[53] Second, he proposed that the speed of light does not change, allowing us to measure intervals between events reliably. From these postulates came an epochal redefining of time, space, and measurement, along with famously surprising insights into strangely behaving clocks, slowed time, twins who live faster or slower, and the equivalence of mass and energy: $e=mc^2$.

Special relativity, however, is limited to uniform speed, which is partly why railroad trains or spaceships are handy examples: They are man-made objects whose speed can be precisely controlled. But Einstein worried: "What has nature to do with *our* coordinate systems and their state of motion?"[54] Indeed, as objects move through the universe, they "fall" and thus accelerate — and vice versa: The two motions are really equivalent. Once acceleration appears, so does gravity. Everything that falls also accelerates thirty-two feet per second during each second traversed. But since special relativity says nothing about acceleration, it also says nothing about gravity, and thus applies only when gravity is absent or negligible (as in subatomic dimensions, which is why atomic physics like Dirac's equation deals only with special relativity).

It is important to note that special relativity remains within the bounds of Euclidean geometry — as did all physics before Einstein. General relativity changed all that in 1915. Euclidean geometry is not only flat (recall high school math), it is also empty; it can hardly describe how monstrous caldrons like our sun pour such fiery energy around them that their gravitational fields skew nearby space. Einstein needed different geometry to describe a universe of energetic intensities and distended curvatures that seem positively surreal next to Newton's clockwork universe. Unlike Euclidean geometry's rigid structures, Einstein's geometry had to bend, flex, or "dimple" according to the energy or mass within it, a geometry not a backdrop for events, but actively part of the events it measures.

In Riemann's non-Euclidean geometry, Einstein found just what he needed. He adapted it to measure how huge masses mold curvatures in space through which lesser bodies "fall." Here, bodies are not pushed or pulled by an outside force, as in Newtonian gravity; they simply "fall" in as straight a line as curved space-time allows — a geodesic, akin to a great circle drawn on the earth's globe.

But general relativity in turn was limited to bodies large enough to feel the power of gravitation. What of the subatomic world? Einstein's inevitable next move was to try joining gravity to the electromagnetic force which reigns in the subatomic dimension. Soon, he took his first stab at a unified field theory.

On July 15, 1925, the German quantum physicist Max Born wrote to Einstein: "I am tremendously pleased with your view that the unification of gravitation with electrodynamics has at long last been successful."[55] Decades later, Born mused: "In those days we all thought that his objective . . . was attainable." Born soon came to believe that Einstein's search was "a tragic error."[56]

Einstein's early attempt assumed that there were two fundamental forces: gravitation, which assemble the planets and galaxies, and electromagnetism, from which all matter is built. (Remember that we now know there are two additional forces.) How do these forces compare? One aspect is their relative strength. Gravity is a very weak force, but gathers strength as it deploys across the vastness of space.[57] The more matter it attracts, the more cumulatively powerful it becomes — until finally it can gather together and swing around the very galaxies. The electromagnetic force is enormously stronger than gravity, by a factor of ten followed by forty-two zeros — luckily so, since the electromagnetic force binds electrons to the nucleus.[58] If gravitational force were stronger in relation to electromagnetic force, matter might dissolve away, and us with it.[59] Further, the strong electromagnetic force prevents negatively charged electrons from repelling each other so violently that they tear atoms and all matter to bits. The electromagnetic force subdues the anarchic tendencies of matter and brings stability to the atomic dimension. Thus can we bathe in waves and particles, light and radioactivity.

How these fields interact was a puzzle, but they must be related; it would be a strange universe otherwise, thought Einstein. If they differ in strength, they are similar in how they work. Both fields are generated by an excitation of matter: gravitation from excited mass (or energy), the electromagnetic field from excited electric charges. It would make sense in the process of unifying if we were to fit such hints and patterns together, as when looking for family resemblances by a common bulk of chin or curve of lip. Einstein had to match mathematical shapes or quirks of matter suggesting kinship: The bones of an equation about gravity might resemble one about electromagnetism, a frequency in an equation about electromagnetism might seem like an oscillation in one about gravity.

Yet the clues led to more puzzles. Thus, the electron's charge might play the same role in electromagnetism that mass does in gravity. However, relativity proved that mass varies with velocity, whereas electromagnetic charge never changes (charge is "conserved"). Even Einstein's success with gravity hindered as much as it helped. Having geometrized gravitation, he now sought an even more general geometry that, while fitting gravity, would include electromagnetism as well. Riemann's geometry worked beautifully for enormous bodies, but could not be applied to atomic phenomena. The charged electron seemed to need a very different geometry than gravity's mutable, swaying geometry — but what sort? Electrons do not really "orbit" the nucleus, as in popular imagination: They move in no usual spatial sense, but rather "up" and "down" in energy levels (quantum "leaps"). They are at once particle and wave, they spin (but only at fixed, quantized levels) and have angular momentum — and this only begins to broach the difficulties.

Einstein had two choices. He could keep the Riemannian framework, but expand the number of dimensions to five or more. Or he could keep the four dimensions, but find a substitute for Riemann's geometry. At one time or another, decade after decade, he pursued both of these possibilities. He explored four- and five-dimensional continuums, differential geometries, gauge transformations, absolute parallelism. He took apart his final field equations for general relativity, assigning the symmetric part to gravitation and the antisymmetric part to the electromagnetic field. He spent the years puzzling at chalk marks on the blackboard.

In 1925, when Einstein set aside his other work to find a unified theory, all seemed in place for success. General relativity provided the geometrizing approach he meant to extend to electromagnetism; he had been mulling the problem for seven years, since 1918; no physicist alive had a deeper intuition of what was physically possible or necessary, or the limits within which the new theory must work. He was still in his prime at forty-six. Max Born, no soft touch, predicted that "physics will be done in six months."[60]

But unified theory was not to become another stroke of genius and insight. It was more like an aging engine fitfully turning over. Periodically, he would declare victory. After his first serious attempt in 1925, Einstein said: "I believe I have found the true solution."[61] He soon decided that he was wrong. He tried other approaches in 1927, and again in 1928 — the latter broadcast as a victory by newspapers to eager readers. Einstein had to dampen the enthusiasm. In 1929, he again believed that he had prevailed, and even gave lectures in France and England. He retracted in 1931. In 1945, at age sixty-six, he published his final equations, but hardly with the overwhelming confidence he had expressed about general relativity. When queried by reporters, he said, "Come back and see me in twenty years." He revised the equations in 1949 and 1954.[62] Always, colleagues and friends bewailed his efforts. In 1932, Pauli was already complaining that Einstein's

> never-failing inventiveness as well as his tenacious energy in the pursuit of [unification] guarantees us in recent years, on the average, one theory per annum.[63]

Einstein himself wondered how definitive even his final equations were. Perhaps, he said wryly, his critics were right, and the equations did not "correspond to nature" — the ultimate defeat.[64] Pauli talked to him in 1954, and said that Einstein admitted with his old directness and honesty, that he had not succeeded in proving the possibility of a pure field theory of matter. He regarded the problem as undecided.[65]

His final equations, appearing at last in 1945, had a muted reception. That year, the atom bombs exploded, the war ended, and physicists sought jobs in the booming field of particle physics. Einstein had become world famous for general relativity. As for unified field theory, it was beyond the horizon. He continued to work on unified field theory until the day he died.

Most physicists now see Einstein's theory as an intellectual feat but irrelevant to physics. A good theory should be able to predict important new insights and express earlier ones in some fertile new way, as general relativity did with Newton. Einstein's attempts do neither. Nor could he take into account the strong and weak nuclear forces. Today, the main contender for unifying all four forces is superstring theory. Most of its proponents pay homage to Einstein as a man "ahead of his time." Still, his disdain for quantum mechanics might well have distanced him from today's unifiers. He tried to circumvent quantum physics by geometrizing electromagnetism in gravitation's image. String theorists have taken the inverse route by quantizing gravitation.

Meanwhile, Einstein's theory exists as a historical artifact in a scientific limbo. Physics does not linger over might-have-beens or maybes, unless they promise discoveries and insights. Careers are short, and the mainstream is where working theories are usually found. Einstein redirected the mainstream in his early work. But his attempts at a unified field theory banished him into the hinterland.

Postscript: The philosopher of science and theology Stanley Jaki, writing of current attempts to unify the forces, suggests that Gödel's theorem might apply. If so, says Jaki, such unification is impossible. Any consistent system, Gödel reminds us, cannot be complete in its own terms.[66] As Einstein strolled to his Princeton office with his friend Gödel, might they have mused about such a limitation on our knowledge of the world?

THE PERSISTENCE OF NATURE

What might Russell have gleaned from his Princeton afternoons at Einstein's house? Much more than he let on, perhaps. Pauli and Gödel were quite biased in favor of metaphysics — as we know, their interests extended to archetypal mythology (Pauli) and the paranormal (Gödel). In a 1946 letter to his colleague Markus Fierz, Pauli spoke of "the idea of the reality of the symbol."[67] Psychology was his link to the "real" world of symbols. It would be difficult to imagine "premises" more distant from Russell's.

Yet Russell's quarrel with Einstein must have been by far the richest and the most important. In a "Note on Non-Demonstrative Inference and Induction," which Russell dictated to his wife in 1959, he offers a tantalizing clue: "My beliefs about induction underwent important modifications in the year 1944, chiefly owing to the discovery that induction used without common sense leads more often to false conclusions than to true ones." He goes on to distinguish between pure induction and what he calls "scientific common sense." Induction, indeed, does not figure in the "extra-logical postulates" required by scientific inference. Here, Russell shows himself to be an empiricist (as ever) with a difference: Induction is "invalid as a logical principle" because it so easily falls into fallacy. Russell's examples of induction going wrong include the following: "No man alive has died, therefore probably all men alive are immortal."[68]

Had Russell in fact given up empiricism, Einstein would have been delighted. In his contribution to the Library of Living Philosophers volume on Russell, published in 1944, Einstein bluntly objected to the Humean "fear of metaphysics" in Russell. Of course, Einstein was quite right. After abandoning Plato in his youth, Russell never let go of the empirical impulse. In a way, the gulf between Russell and Einstein was not enormous. Neither sub-scribed to what Einstein called the two illusions: "the unlimited penetrative power of thought" and "naïve realism, according to which things 'are' as they are perceived through our senses."[69] Ein-stein agreed that "thought acquires material content only through its relationship with . . . sensory material" — a statement that sounds suspiciously acquiescent to empiricism. But he rejected any attempt to base thought upon material reality, arguing that the "free creations of thought" are sufficiently valid if they are merely "con-nected with sensory experiences." That is, thought is not created out of material things or the perceptions of material things. But thought can contribute to knowledge only if it coincides with the "sense experience" that comes to us from what is material. Einstein was, to borrow his own words, on the thought side of the "gulf — logically unbridgeable — which separates the world of sensory ex-periences from the world of concepts and propositions."[70]

The format of each volume of The Library of Living Philoso-phers requires that the subject "reply" to each essay. Russell duti-fully replied to Einstein's contribution. Russell's few words are respectful and pointed. He agreed with Einstein that the "fear of metaphysics is the contemporary malady" — lamentable especi-ally for the tendency of contemporary philosophers to swallow empiricism wholesale, without "prob[ing] questions to the bot-tom." Still, Russell approached the "gulf" between metaphysics and empiricism with a "bias . . . towards empiricism." He is thus quick to refute Einstein's assertion that number is an example of the "free creations of thought."[71] As one contrary instance, Russell offered the obvious correlation between the decimal system and our ten fingers. For Einstein, desperately clinging to the hope of a mathe-matical solution to his unified field theory, Russell's empirical bent must have been an unpleasant reminder of those abandoned "gen-eralizable facts" upon which his relativity theories were based.[72]

In 1949, Russell wrote "Einstein and the Theory of Relativity" for a BBC broadcast. In it, Russell praised modern physics for its "desire to avoid introducing into physics anything that, by its very nature, must be unobservable." The consequence has been more abstraction in physics, as no longer are we permitted "to make pictures to ourselves of what goes on in atoms, or indeed of anything in the physical world." The tongue-in-cheek of this quip aside, Russell put his finger on the paradox of evidence in physics. What Russell wanted was less of the unobservable to count as science: "[S]o long as the technique of science can survive, every diminution in the number of unobservables that are assumed is a gain. In this sense, Einstein took a long step forward." [73] Russell had recast Occam's razor* to fit modern physics, but how that law of parsimony can be reconciled with scientific creativity, much less a "theory of everything," is hard to know.

If Einstein did waver in his commitment to experience, he never fully gave it up. Asked by *Scientific American* to explain his most recent unified field theory in nontechnical terms, Einstein obliged. The result, a difficult and abstract article published in 1950, conceded as much:

> The skeptic will say, "It may well be true that this system of equations is reasonable from a logical standpoint. But this does not prove that it corresponds to nature." You are right, dear skeptic. Experience alone can decide on truth.[74]

---

*William of Occam admonished, "Pluralitas non est ponenda sine necessitate," which can be translated, loosely, as "Strip away unnecessary things."

PART 4

BEYOND PATHOS: OPPENHEIMER, HEISENBERG, AND THE WAR

As Einstein and his friends in Princeton spoke quietly of philosophy and science, many of their colleagues were busy pushing physics toward brute power. In Los Alamos and in Germany, physicists raced to build the first atom bomb. Whoever succeeded would gain certain victory: The heart of London or Berlin could be destroyed in a moment. Neither Einstein nor his fellow physicist Pauli worked on the atom bomb, but both knew what their colleagues were doing. It was a small, tight-knit world. Oppenheimer, who directed the Los Alamos effort, had been Pauli's student. Werner Heisenberg, who led the German bomb project, was Pauli's closest collaborator. They had been friends since their college days in Munich. When it came down to the atom, everyone knew everyone else.

WARTIME BERLIN, WINTER 1943–44

IN DECEMBER 1943, WERNER HEISENBERG paid a visit to Krakow at the invitation of Hans Frank, then the Nazi governor general of occupied Poland.[1] Frank, a schoolmate of Heisenberg's brother, had extended the invitation in May 1943, having invented a "Copernicus

Prize" for Heisenberg as an enticement. Delayed for various reasons, Heisenberg finally accepted, with the promise, too, of a lecture for Frank's newly minted Institut fur Deutsche Ostarbeit (literally, Institute for German East Studies, a "think tank" for eastern colonization).

Frank had fallen from Nazi grace the previous year after a lecture critical of unconstitutional rule. That lecture saw him stripped of his prestigious title Reichskommissar, but left his governorship of Poland intact. In such disdain did Hitler hold Poland that he thought it punishment enough to let Frank languish there.

Exiled he might have been. But he did not languish. He had already appropriated for his living quarters the luxuriously appointed Wawel Castle, where he entertained lavishly and famously. Known as the "butcher of Poles" (later executed at Nuremburg in 1946), he extorted from his governorship all that he could: lavish feasts for friends, furs for his wife and his lover, money in the bank. (Among the charges of corruption floated during party infighting was the charge, easily documented, that Frank and his wife "shopped" in the Jewish ghetto, where discounts naturally abounded.) Frank despised Poles, whom he saw as fodder for slavery and extermination; and he exhorted his fellow Germans to exterminate Jews in a blunt, brutal 1941 speech.

In recollection, Heisenberg confided to the historian David Irving that he was struck by Frank's queries about a "miracle weapon, perhaps atomic bombs" in the possession of the Allies.[2] Heisenberg seems to have recalled little about his own talk at Frank's Institut fur Deutsche Ostarbeit. Presumably, he lectured on quantum theory. Nor did he recall seeing or hearing anything untoward. Yet he must have listened to the outspoken Frank boast of his successes in dealing with the "Jewish question." Frank's castle was about fifty miles from Auschwitz.

It was hardly surprising that Hans Frank might think to ask about an atomic bomb. Whether he had in mind a "miracle" weapon for the fatherland or feared that the Allies might have their own, Frank, like most laypeople, would have heard all about "splitting the atom" and the possibilities of atomic energy.

Indeed, by the late 1930s, most physicists were at least speculating on the possibility of creating an atomic explosion. The idea was the logical outgrowth of three decades of revolutionary thoughts about the forces of nature, both large and small. Einstein's theories of relativity had recast how we see the large forces of the universe — gravity, the speed of light. The smallest forces — those within the atom — were next. By 1912, the British physicist Ernest Rutherford had suggested a model for the atom. He filled his laboratory in Manchester with eager young physicists. One, Niels Bohr, emerged as the single greatest theorist of the quantum. When he established his own laboratory in Copenhagen, he attracted a cadre of youngsters eager to take on the atom and make their own marks in history. Among Bohr's students was the young Bavarian Werner Heisenberg.

They met at a lecture given by Bohr in 1922 (the year Bohr won the Nobel Prize). At once, the twenty-year-old Heisenberg impressed Bohr. His questions were pointed and probing, he was not afraid to argue, and he possessed great energy. In the following few years, working with fellow Germans Wolfgang Pauli, Max Born, and Pascual Jordan, Heisenberg developed the foundations of quantum mechanics. From the discovery of "matrix mechanics" to his famous "uncertainty principle," he played a part second only to Bohr's in the story of the quantum.

HEISENBERG

Born in 1901 in Würtzburg, a languid and venerable Bavarian town, Heisenberg was raised in a typical patriarchal family. His father, August, was a Greek scholar, ambitious and successful — he passed his "habilitation" and became Professor of Middle and Modern

Greek in 1909, when Werner was eight. The following year, the family moved to Munich, where Werner entered the Maximilians-gymnasium, an illustrious and traditional school where students received instruction in Latin, Greek, mathematics, and, to a much lesser extent, subjects such as history, geography, and athletics. Physics was an afterthought. Werner quickly established himself as a star in mathematics.

August Heisenberg was ambitious not only for himself but for his sons as well. His desire for hearty sons was manifest in frequent "tests" pitting one son against another. These games probably contributed to Werner's obsessive competitiveness. They also bred extreme antipathy: By mid-adolescence, the brothers were fighting bitterly. Their relationship remained frosty at best throughout their lives.

When World War I intervened, not only was Werner's schooling transformed, so too was his home life. His father, an army officer, was immediately called to active duty. He served enthusiastically, eventually volunteering for the front. Within two weeks, his naïve patriotism was tempered by the "pain, misery and suffering" he witnessed.[3] He requested a transfer back to Munich in April, leaving the young men in his regiment to fight an old man's war.

Transformed, too, was Werner's educational experience. A new building built for the Gymnasium was turned over to the military. Some faculty went off to the war, only to return quickly, as had Werner's father. Of the seventy-four young students who enlisted, more than one-third were killed. The Gymnasium, in addition to supplying cannon fodder in the form of its pupils, exhorted the younger students to displays of patriotism in support of the war. Werner joined the "Military Preparedness Association," a national organization with chapters at each Gymnasium. Had the war continued, Werner, who turned seventeen in December 1918, would undoubtedly have served.

One interlude during the war may have changed Heisenberg

profoundly. In 1917, he spent a long summer in the countryside, where he and other students joined in harvesting much-needed hay. There, imbued with the romanticism of hard labor, he studied mathematics and played chess.

In 1920, Heisenberg began studies at the University of Munich. He dazzled his professors, publishing important papers on atomic structure while still a fledgling student. His early love of mathematics was about to pay off. During his last years at the Gymnasium, he had worked through the mathematics of general relativity. His conversion to physics came late in his Gymnasium studies.

All around him, Germany was in chaos. Steeped in the elitist politics of his upper-middle-class academic family, Heisenberg remained with the patriotic Military Preparedness Association, renamed the Young Bavaria League. It encouraged near-cultlike "retreats" into the countryside, where leaders like Heisenberg conducted seminars on truth, honesty, and the cleansing power of nature. Nominally apolitical, the group offered a romantic alternative to the difficult politics of the Weimar Republic. Throughout the early 1920s, groups like Heisenberg's habitually broke off from one organization and joined another. Heisenberg's group seems to have resisted joining any of the more virulent anti-Semitic organizations and retained its devotion to nature and traditional values. Still, he was drawn to science as a transforming enterprise, hoping to work in "those fields in which it was not simply a question of the further development of what is already known."[4]

The early 1920s brought about such transformation in our knowledge of the atom that the world still reels from the impact. Heisenberg and his fellow student–colleague Pauli joined Arnold Sommerfeld's Theoretical Physics Institute in Munich. Soon, Heisenberg and Pauli were collaborating with Niels Bohr in Copenhagen and Max Born in Göttingen on what would become the new quantum physics. It was Heisenberg — sometimes with Bohr's approbation, sometimes without — who forged the beginnings of quantum mechanics and hit upon the uncertainty principle. With Bohr and Pauli as sounding boards and critics, Heisenberg blossomed:

> The five years following the Solvay Congress in Brussels looked so wonderful that we often spoke of them as the golden age of atomic physics. The great obstacles that had occupied all our efforts in the preceding years had been cleared out of the way; the gate to that entirely new field — the quantum mechanics of the atomic shell — stood wide-open, and fresh fruits seemed ready for the plucking.[5]

In 1932, Heisenberg won the Nobel Prize in Physics for the "creation of quantum mechanics, the application of which has led, among other things, to the discovery of the allotropic forms of hydrogen."

The golden age came to an abrupt end in 1933. The Nazis quickly set about cleansing the German civil service — including the universities — of Jews and other misfits. Aryanism became the modus operandi of state security. Who could be trusted to serve the Nazi regime?[6] Aryanism supplied the answer: Only those of the correct race could be so trusted. Serving Hitler's State demanded absolute obedience and sacrifice. Anyone with Jewish blood was perforce untrustworthy.

In 1934, this covenant was backed up by a new law requiring all civil servants to swear personal allegiance to Hitler:

> I swear that I will be loyal and obedient to the *Führer* of the German Reich and people, Adolf Hitler, respect the laws, and exercise the obligations of my office conscientiously, so help me God.[7]

As Mark Walker notes, the oath was administered with a rather elaborate ritual, to reinforce the point: The administrative head read the oath aloud, the others repeated it in unison, and each confirmed it in writing. Those who refused were fired.[8] Heisenberg and his fellow physicists, as employees of the state, swore allegiance.

One of those happy to take the oath was Otto Hahn. Working with Lise Meitner, Hahn spent 1939 in a laboratory at the Kaiser Wilhelm Institute aiming neutrons at nuclei. One day, he aimed a neutron squarely at a uranium nucleus and was astonished to find that it split in two. He had discovered nuclear fission. It took only a few months for word to spread across the globe. Scientists from Britain, Germany, the Soviet Union, and Japan debated the technicalities of an atom bomb. When Germany invaded Poland in 1939, Heisenberg joined what would become known as Hitler's Uranium Club. Soon nine "atomic" task forces, coordinated by Kurt Diebner, were at work in Germany. Heisenberg traveled back and forth from Leipzig to Berlin, busily supervising efforts in those cities. The German atomic effort was under way fully three years before the United States launched its Manhattan Project.

If the Germans lost the race to build the bomb, they certainly succeeded in confounding historians. Documents recently unearthed in Russian archives suggest that Diebner's group might have succeeded in setting off a small thermonuclear bomb in 1945, killing a number of slave laborers.[9] Heisenberg's role is by far the most hotly disputed. Did he subvert the Nazi atomic effort, as he claimed, or did he simply bungle the mathematics? Michael Frayn's play *Copenhagen* has rekindled the debate, dramatizing but never clarifying Heisenberg's 1941 meeting with Niels Bohr in Copenhagen.

What were Heisenberg's motives? Perhaps Heisenberg himself never knew. However pure his devotion to science, it could not withstand his patriotism. He was the most prominent physicist to remain in Germany under the Nazi regime. Yet in 1936, he found himself under attack as a "white Jew" — a colleague and friend to the "Jewish physicists" Bohr, Pauli, and Born. He was derided in Nazi publications and reproached for mentioning Einstein when he taught relativity theory.[10] Desperate, Heisenberg used his family connections. Through them, he appealed to SS leader Heinrich Himmler to clear his name. The accusations melted away. A revised security report read: "Heisenberg's political position is in no way to be designated argumentative. He is undoubtedly the unpolitical academic type."[11]

So Heisenberg remained in Germany, a scholar immune from politics. Paul Lawrence Rose puts it succinctly: For those like Heisenberg, "their support was spiritual, patriotic, social, national, cultural, moral, natural — indeed anything but the detested 'political' behavior that defense of the Weimar Republic represented."[12] In Heisenberg's mind, military service was outside politics, and, an avid outdoorsman, he took to it happily. At eighteen, he volunteered for a cavalry unit aiming to fight Communists in Munich. "These two years had tremendous significance for my human development. My position on political questions was perhaps then decided."[13] He remained active through the mid-1930s, training periodically in an army reserve unit.

Scientists like Einstein might stoop to politics. Even his friend Max Planck thought Einstein's pacifism and Zionism too "political." For Heisenberg, physics, pure in its nobility, lay outside the degradations of politics.[14] Ironically, by working on atomic fission, he could defend the fatherland, yet remain unsullied by Nazi policy.

In truth, German physics had become fully Nazified. The Deutsche Physik (German Physics) movement boasted two Nobel laureates as members, Philipp Lenard and Johannes Stark. German Physics would defend Nazi ideology, the cult of the Führer, and the wish for a return to a mythical German past. If modern physics endorsed the liberal, democratic, international world, German Physics would fight every vestige of "modern" decadence. If modern physics represented rationalism, German Physics would find ways around the influence of "subversive" scientists, appealing to "will" over the claims of reason.

But this scientific crusade inevitably collided with economic and military realities. German industrialization depended on high technology, and thus on fundamental work in physics. An army of skilled engineers and technicians had built the German military machine, and skilled engineers would be required to keep it running. One theory as to why the Germans failed to build a bomb is simply that there seemed no need. Until Stalingrad in 1942, the march across Europe had been victorious. Once Stalingrad was lost, time had run out. Heisenberg and his nuclear team lacked the resources of uranium and heavy water needed to make a bomb. Even counting several teams at work, the German effort never employed more than several hundred. Tens of thousands worked on the American bomb.[15]

At the end of the war, an American team captured Heisenberg, Otto Hahn, Carl Friedrich von Weizsacker, Max von Laue, and six other prominent physicists. They were sent to Britain for direct interrogation and for what could be learned by eavesdropping. The secret taping was yet another unprecedented step taken in the name of "security." The greatest fear was that the fruits of German science might fall into Soviet hands. From the Allied viewpoint, it was vital to find out what Heisenberg and his colleagues knew and when they knew it. As it happened, the Germans seem to have failed. But whether Heisenberg had the knowledge and refrained from using it, or had, in fact, made a fatal calculation of the uranium required, it took him very little time (about twenty-four hours) to come up with an explanation of how the Americans had split the atom and created the atomic age.

Thus, at a house called Farm Hall in the verdant countryside of England, near Cambridge, Heisenberg and his colleagues spent six months talking to one another and, periodically, to the British officers who politely "detained" them. Farm Hall, manorial though it seemed, had been a "safe house" for MI5. It was outfitted with microphones. From July 1945 through early January, the German physicists were surreptitiously taped. The possibility of eavesdropping did not seem to have occurred to the Germans, who spoke freely among themselves. Indeed, never before had nations taken such scientific hostages.

Once released, Heisenberg came home a hero to his countrymen. He had failed to build the bomb that could have gained Germany's victory, but no disgrace awaited him. Whether true or not, he claimed to have foiled the German bomb effort by deliberately slowing it. His claim seemed to take the high moral ground compared to the United States' atrocity against Hiroshima. The Germans could not escape the grisly fact of their death camps, but Heisenberg made it possible for them to look down on those who had so dramatically perverted the purity of science.

Whether Heisenberg intended to fail or not, we do not know. But to those in Los Alamos, Heisenberg as director of the German effort seemed a distinct threat.

WARTIME LOS ALAMOS, WINTER 1943–44

On October 3, 1943, Niels Bohr, the eminent and avuncular "father" of quantum mechanics, donned an ill-fitting aviator's helmet and squeezed into the bomb bay of a twin-engine plane used by England to ferry diplomatic pouches from Stockholm.

Bohr had escaped just days before from his homeland, occupied Denmark, which had become too dangerous for Jews. Until 1943, German Nazis had reluctantly acceded to the Danish monarchy and populace. In return for Denmark's agricultural riches, the occupiers had refrained from their usual tactics. No longer. The tide of the war was beginning to change against Germany, and as it contracted, the pace of the Holocaust quickened. Berlin ordered the arrest of Bohr and his brother Harold, to be followed by the arrest of all Danish Jews. Tipped off, Bohr and his wife casually walked out their door and down to the seaside, where they hid until, under cover of night, they were ferried across the Oresund to Sweden.

Even Sweden was too dangerous. Its streets were crawling with German agents ready to bundle Bohr back to Germany. Intent on brokering a deal to save the Jews of Denmark, Bohr rushed (incognito) to Stockholm, where he appealed to King Gustaf V. At first, the king demurred, having been rebuffed by the Nazis when, in 1940, he offered asylum to the Jews of Norway. Bohr persisted, and with Swedish help, the Danish underground managed to ferry some six thousand refugees (virtually the entire Jewish population of Denmark) across the Oresund.

Within days, Bohr received an official invitation to join fellow scientists in Britain. Certain that his family was safe in Sweden, he accepted. Again, the route out was hazardous. Bohr huddled in the empty bomb bay of a two-engine Mosquito used for diplomatic mail. He was unconscious from oxygen deprivation by the time he arrived in London, but quickly recovered. From there, with his son Aage, he soon traveled with the British contingent to Los Alamos, New Mexico, where J. Robert Oppenheimer and his colleagues were constructing their "completely fantastic" gadget.

Los Alamos was the nerve center of an effort to devise the most destructive "gadget" ever imagined. There, theorists and engineers worked on the explosive mechanisms. Two other locations — Oak Ridge, Tennessee, and Hartford, Washington — housed vast facilities for producing fissionable material. Additional work took place in university laboratories across the country. It was, of course, a secret mission, though its vast size alone made security nearly impossible.

Bohr traveled to Los Alamos by train, along with his son and General Leslie Groves, commanding officer of the Manhattan Project. If Bohr was astonished by the progress in atomic fission, he must have been stupefied by the immensity of the effort. Like Oak Ridge and Hartford, Los Alamos was a small town. In a matter of months, it had grown from a tiny outpost of a few school buildings into a miniature city of tenement apartments, dormitories, mess halls, physics and chemistry laboratories, warehouses, medical offices, a machine shop, a school, and even a radio station.

The whirlwind construction was overseen by the yin and yang of the American atomic bomb effort: General Groves and J. Robert Oppenheimer. A formidable commander with an ego to match, Groves had trained as an engineer at MIT and West Point, where he graduated fourth in his class. He was aggressive, blustery, and dictatorial — not the sort of man who might endear himself to experimental and theoretical physicists. Indeed, his whirlwind tour of the project laboratories in September 1942, beginning with Pittsburgh and New York and heading west, left egos bruised and Groves increasingly distressed. He was visibly irritated by the prevailing academicism — open-mindedness was valued above efficiency, self-doubt above ambition. "When in doubt, act" was his dictum and constant theme. "How can you work with people like that?" asked Leo Szilard of his fellow physicists in the Chicago Metallurgy Laboratory.[16]

By the time he reached Berkeley, Groves was desperate for someone who could offer clarity and certitude. What he found was Oppenheimer — his polar opposite in character and intellect. The bookish, leftist, artistic, rail-thin theorist might well have come from a different universe. But the two men had one thing in common: a facility for administration and organization. Oppenheimer had joined the nuclear weapons project officially in February 1942; by May, he was heading various teams at work on the construction of the bomb mechanism. By October, when Groves walked into his office, Oppenheimer and his groups had made great progress in determining the scale, structure, and impact of the bomb and the fission process.

Impressed with what was, for once, a clear and realistic report, Groves invited Oppenheimer to accompany him on the leg back from Chicago. In a small train compartment, Groves, two of his aides, and Oppenheimer dreamed up what was to become Los Alamos: not a cloistered university for scientific speculation and dialogue (or so Groves would have it), but a single, unified, purposeful laboratory where scientists could exchange ideas freely, without security concerns. For Oppenheimer, the idea of a central laboratory must have seemed quite natural, given the communitarian nature of science. For Groves, rounding up all of those scientists into one place would at least facilitate scrutiny, if not discipline. Oppenheimer had in mind an isolated spot northwest of Santa Fe, New Mexico. For Oppenheimer, it was familiar ground. He had spent vacations horseback riding and hiking through the desert land. He and Groves traveled to New Mexico to find the best site. The first possibility, deep in a valley, lacked not only sunlight but also even the hint of an infrastructure. The second, Oppenheimer's suggestion, came with buildings and infrastructure, in the form of the Los Alamos Boys School, and was high on a plateau — secluded and serene.

Today, such a project might be deemed a boondoggle, a waste of taxpayers' monies. There was no assurance that the bomb could be built, and the expense of building not only laboratories, but the small town that served as living quarters, was enormous.

Groves controlled the purse strings, which he held tight whenever it came to making the town more livable — intersections lacked traffic lights, dwellings consisted of barracks in army issue green, garbage pickups were uncertain. Still, wrote Ruth Marshak, wife of a Los Alamos scientist, the barren landscape had its virtue:

> As we neared the top of the mesa, the view was breathtaking. Behind us lay the Sangre de Cristo Mountains, at sunset bathed in changing waves of color — scarlets and lavenders. Below was the desert with is flatness broken by majestic palisades that seemed like the ruined cathedrals and palaces of some old, great, vanished race. Ahead was Los Alamos, and beyond the flat plateau on which it sat was its backdrop, the Jemez Mountain Range. Whenever things went wrong at Los Alamos, and there was never a day when they didn't, we had this one consolation — we had a view.[17]

However difficult it proved to build and supply a major laboratory in this isolated desert Eden, Los Alamos satisfied Oppenheimer's craving for beauty and Grove's for security.

From a few school buildings, Oppenheimer and Groves created a bustling, makeshift town with a hospital, a school, a town council, ceaseless shortages of housing and water, and a mania for security. Some four thousand civilians and two thousand military personnel were working at Los Alamos by war's end. Those who came in the early days of spring 1943 endured countless deprivations: The laboratories were works in progress, the prefabricated housing turned shabby directly upon being erected, roads were dusty and haphazardly graded, and mail and the rare telephone call were censored.

Still, Oppenheimer managed to recruit, either as residents or as consultants, dozens upon dozens of first-rate scientists: Edward Teller, I. I. Rabi, Enrico Fermi, Ernest Lawrence, Emilio Segré, Hans Bethe, Otto Frisch, Niels Bohr, John von Neumann, and a host of younger men who would become famous, among them Richard Feynman and Robert Wilson. Physically isolated and deprived of cultural distractions (save intense partying and dancing on the weekends), Oppenheimer's scientists worked hard from eight-thirty in the morning until well into the night.

By the winter of 1943, the project moved from theory to construction. Robert Wilson's group had established the "critical mass" of uranium 235 — that is, the amount of this type of uranium necessary to create a chain reaction. All that remained was to create sufficient fissionable material, figure out how join two quantities together to create the chain reaction, and test the product. The bomb makers turned from theoretical science to engineering.

Two problems were paramount: manufacturing sufficient fissionable material and creating a mechanism to explode the material. Manufacturing fissionable material proved to be tricky at best. At Tennessee's Oak Ridge, two plants — one using electromagnetic force, the other gas diffusion — were under construction by the summer of 1943. The first was operational by the fall, but needed constant repairs. The second remained unfinished (and its technology still uncertain) by the year's end.

Constructing an explosive mechanism proved trickier still. Nuclear fission, unlike TNT or other conventional explosives, must begin with sufficient material to achieve critical mass. Of course, the atomic material must remain "subcritical" until the moment of explosion. By the end of 1943, the possibilities had narrowed to two: a "gun" method, whereby an amount of subcritical material is shot into another, producing a supercritical reaction; and an "implosion" method, whereby an amount of subcritical material is imploded into itself by surrounding explosives, thus compressing the fissionable material and turning it supercritical. It would take more than a year to engineer working explosive devices.

Thus, during the winter of 1943–44, Oppenheimer juggled numerous teams at work on designing and arming an atomic bomb. He was seldom away from Los Alamos. Although fond of horseback riding, he rarely took part in the town's social scene.

Bohr arrived at Los Alamos in December 1943. He came with a very tantalizing sketch of a heavy-water reactor given him by Heisenberg during their famous 1941 conversation. Astonished at the successes of the Manhattan Project, Bohr now found his early skepticism about atomic fission evaporating. Could Heisenberg's boxlike drawing be the prototype of a bomb?

A roundtable of Los Alamos physicists pondered the sketch and came to the conclusion that "these Germans were totally crazy," in the words of Hans Bethe.[18] Was this reactor an earnest effort at a nuclear bomb? If so, the Los Alamos scientists could only scratch their heads in wonderment at what was obviously an inefficient and ineffective device. Or was it a smoke screen, intended to lull the Allies, via Bohr, into a false sense of security over Germany's atomic progress?

We now know that Germany, having persecuted and expelled its greatest physicists, lagged far behind in the race to produce a bomb. But at the time, Heisenberg's sketch wasn't enough to settle the matter. Indeed, if the Americans had progressed so far, was it not possible that German physicists, Heisenberg among them, had made equal progress? Thus was born Alsos, an intelligence gathering project aimed at finding out whether Germany might be on the brink of developing an atomic bomb.

Aside from the dubious sketch, Bohr brought to Los Alamos fatherly approbation and a sense of mission. By the winter of 1943, the project had been fully theorized and seemed on its way to certain completion, depending only on the solution to various technical problems. Owing, perhaps, to this seeming confidence, the Los Alamos scientists began to ponder the ethics of their project. Oppenheimer must have voiced his anxieties to Bohr, for he later recalled Bohr's "having made the enterprise hopeful, when many were not free of misgiving."[19] Bohr had little to add to the technical side of the bomb. But he was an éminence grise whose humanity was legendary. He was also an emblem of the rape of Europe. His escape from Hitler's grasp, so dramatic and timely, galvanized the men and women cloistered in the mountains of New Mexico. Later, Oppenheimer remembered Bohr as having spoken of Nazi oppression and of his own "high hope that the outcome would be good."[20]

At Los Alamos, Bohr saw the dawn of the nuclear age. Always prescient, he knew almost instantly that the arms race had begun. His answer: a theory of "political" complementarity. The atom bomb was, he felt, the technology that could make war obsolete — but only if its workings were made transparent. In later, fruitless discussions with Roosevelt and Churchill, he proposed sharing atomic information with the Soviets as a gesture of goodwill and diplomacy. Such openness, he argued, would reduce the chance of a postwar arms race. Roosevelt was reported to have listened intently and sympathetically. Churchill, on the other hand, rudely handed Bohr his hat — and later seems to have persuaded Roosevelt that Bohr and his ideas were dangerously naïve, or simply dangerous.

The danger was yet to come.

OPPENHEIMER

Julius Robert Oppenheimer was born in New York in 1904, the son of an affluent German Jewish couple. Precocious, bookish, adventurous, and high-minded (having been schooled at the New York Society for Ethical Culture), he was also petulant and physically lazy, particularly during his teenage years. At eighteen, having delayed entrance to Harvard for a year owing to digestive problems, he spent several weeks camping in New Mexico and Colorado with Herbert Smith, his high school English teacher. Not only did his health improve, but he also fell in love with the rugged landscape and vast mountains. He returned to college much improved in disposition and health. After sailing through Harvard in three years, summa cum laude, he set off for Cambridge, England, and the great experimenter Ernest Rutherford. But he lacked experimental gifts, and soon shifted to theoretical physics at Göttingen. He took his doctorate — in one year — under Max Born in 1927, and meanwhile made his name by publishing sixteen first-rate papers by 1929.

In 1929, Paul Ehrenfest, Einstein's close friend, arranged for Oppenheimer to study with Wolfgang Pauli. Oppenheimer had brilliant ideas, but was poor at calculations. Pauli, a superb and rigorous calculator, put Oppenheimer through the mill. It says much about physics in the 1920s that Pauli was only four years older than Oppenheimer, yet seemed the master to Oppenheimer's apprentice. What made the difference was the pace of discoveries in physics in that decade. The brunt of that advance was accomplished by 1927 — by then, someone like Oppenheimer could feel as if he had missed the golden age by a twinkling.

He went to Berkeley in 1929 and by the late 1930s had built — as Hans Bethe put it flatly — "the greatest school of theoretical physics that the United States has ever known." This was more than an academic matter. Through the 1920s, the best young American physicists studied in Europe, especially Germany. When Oppenheimer returned to the United States, he made it possible for American students to be educated at home. The arrivals of refugees from Hitlerism only strengthened this achievement. When research for the atom bomb seriously began in 1942, American physics, the equal of any in the world, was ready.

There is no end of testimony to Oppenheimer's brilliance, and justly so. His research on quantum physics made him an international force.[21] His graduate students, fascinated by him, mimicked the way he spoke, smoked, and gestured. He attracted women, daunted colleagues, and by sheer intellectual speed and range overwhelmed many of his peers. The physicist Emilio Segré said Oppenheimer had the quickest mind he had ever seen — no small praise: Segré had been trained by no less than Fermi himself. The young Edward Teller was overpowered by Oppenheimer's mind and personality.[22] To those around him, he glittered.

The inner Oppenheimer was much more complicated and troubled. He had grown up rich, sheltered, and rather spoiled. He could not help using his intelligence to browbeat others. He was feared for his sarcasm, which, unlike Pauli's, seemed personal and even vicious. Certainly, Oppenheimer's taste for humiliating others reflected his own insecurity. He could analyze, criticize, absorb, and penetrate all difficulties with astonishing ease with his "iron memory." But he never succeeded in producing truly creative work. For someone so gifted, it must have been a bitter failure. The psychological burden seems to have lifted when he directed Los Alamos. There, his critical gifts were exactly what was needed, and his restless energy was wholly occupied. In those years, he was self-confident and at ease with himself.

During his early years at Berkeley, Oppenheimer's private interests were as rarefied as his physics. He describes himself then:

> I studied and read Sanskrit with Arthur Ryder. I read very widely, mostly classics, novels, plays and poetry; and I read something of other parts of science. I was not interested in and did not read about economics or politics. I was almost wholly divorced from the contemporary scene in this country. I never read a newspaper or a current magazine like *Time* or *Harper's*; I had no radio, no telephone; I learned of the stock market crash in the fall of 1929 only long after the event; the first time I ever voted was in the Presidential election of 1936. To many of my friends, my indifference to contemporary affairs seemed bizarre, and they often chided me with being too much a highbrow.[23]

In 1936, his interests shifted radically. The exquisite aesthete became a political activist. The transformation was sparked in part by Oppenheimer's growing awareness of Nazi persecutions. It was also encouraged by his affair with the moody, smart, beguiling, and sometimes badly depressed Jean Tatlock. She introduced Oppenheimer to the world of left-wing protest and intrigue, to Communists, union organizers, and Spanish Civil War loyalists. She herself had been a Party member off and on. It was a heady experience for the privileged and precious Oppenheimer. The affair did not last, however. (Four years later, the gifted but troubled Tatlock committed suicide.) In 1940, Oppenheimer married Katherine Puening, whose first husband, Joe Dallet, had died fighting in the Spanish Civil War. Dallet had been a Party organizer; Katherine had been a member herself for several years. By 1939 and the Nazi-Soviet Pact that led to the devouring of Poland, she had become disillusioned. Oppenheimer, too, began to gravitate away from the extreme left.

But his political past returned to haunt Oppenheimer. When General Groves, head of the bomb project, decided that Oppenheimer was the ideal director, he was rebuffed, at first, by the Army. Groves persisted and managed to cow the security-conscious Military Policy Committee.[24] Oppenheimer's organizational brilliance surprised many of his colleagues, who were first dismayed to have a young man who had not even won a Nobel Prize leading Los Alamos. Despite his support of Oppenheimer, Groves was intensely security-minded. So guarded was he of atomic secrets that he hesitated to brief agents sent behind Germany lines lest a captured agent unwittingly give away a vital secret.[25] Still, Groves could not prevent background checks on Oppenheimer, and throughout 1943 they continued. He was shadowed, his Berkeley neighbors were questioned, and he was interrogated endlessly. FBI reports accumulated. Whenever Oppenheimer was questioned, his answers were recorded. Later, careless errors and contradictions were pointed to as evidence of questionable loyalty. In June 1943, Oppenheimer visited Berkeley and then met Jean Tatlock in San Francisco. The FBI trailed them on leaden feet:

> He was met by Jean Tatlock who kissed him. They dined ... then proceeded at 10:50 pm to 1450 Montgomery Street and entered a top-floor apartment. Subsequently the lights were extinguished and Oppenheimer was not observed until 8:30 am next day when he and Jean Tatlock left the building together.[26]

In late 1943, with the supersecret Los Alamos laboratory well under way and awaiting his attention, its director was forced to reveal the name of a friend, Haakon Chevalier, a former colleague at Berkeley. Chevalier had earlier approached Oppenheimer with the name of an engineer who had Russian contacts. Having at first failed to mention the incident to the FBI, he later did so. The admission, and his failure to disclose the conversation immediately, became part of the evidence at his 1954 hearing.

Security dogged Oppenheimer, as it did (and still does) all physical scientists underwritten by a government at war (whether hot or cold, declared or not). In the 1940s, security meant Army G-2 (Intelligence) and the FBI, whose director, J. Edgar Hoover, was rabidly anti-Communist and reflexively anti-Semitic. In the earlier Red Scare of 1919, Attorney General Mitchell Palmer had launched a campaign against "foreign-born subversives and traitors,"[27] by which was meant anyone connected to Marxism, socialism, trade unionism, or a myriad of other left-wing inclinations. (It did Hitler no harm among many conservative Americans that he declared himself an enemy of Bolshevism.) Likewise, Palmer's protégé Hoover believed that the Communist threat came from within. When, later, Oppenheimer faced his inquisition, his work at Los Alamos seemed to constitute "means and opportunity." The motive was a given.

Los Alamos meant the end of Oppenheimer's scientific career. Like all his Los Alamos colleagues, he did no fundamental work during the war. Afterwards, he became a public figure, for better or worse. Yet Los Alamos in turn did much for Oppenheimer. The astute Hans Bethe, who knew him well, later observed:

> There was a tremendous change in Oppenheimer from 1940 to 1942, and especially in 1943. In 1940 he was confused, he mumbled, he certainly wouldn't have given anybody any orders. . . . [H]e was attracted by problems beyond the capacity of anybody to solve, including his. . . . In 1942 the new personality had gelled. He was much more decisive. . . . [In 1943] he really came into his own, and he obviously had always wanted to accomplish something definite, something outstanding. And Berkeley and Caltech had not given him that opportunity.[28]

Inside the esoteric theorist was a man of action, happy to escape into the world. His early political activities had launched him into the arena of action; at Los Alamos, he led. It must have seemed a sublime duty, to ensure the military safety of the United States against the Nazis. He could at once hold his own with the brilliant Fermi and keep in mind every detail from the number and size of the mess halls to the need for code names for Niels Bohr and his son. He bent the rules to bring Feynman's beloved and dying wife to New Mexico. He stood up to his own strong-minded superior, General Groves, who wanted all the physicists commissioned, to keep them under strict Army regulations.

After Hiroshima, Oppenheimer became a national hero. He was the man who had ended the war. His thin, ascetic face was everywhere, in magazines and newspapers and even on TV. To be a "theoretical" physicist suddenly seemed glamorous. Oppenheimer brought his organizational acumen to Princeton as the director of the Institute for Advanced Study. There, he quietly assisted the government in its atomic policy, armed guards watching over a safe near his office.

But even at the height of his power and celebrity, Oppenheimer was vulnerable. With the end of World War II came the Cold War. The Soviet Union, not surprisingly, developed its own bomb (aided by the Los Alamos spy Klaus Fuchs). The House Committee on Un-American Activities (HUAC) began hearings in the late 1940s. Charges began to fly. Communist subversives were said to infest the unions, own Hollywood, and have infiltrated the government. Inevitably, the Committee turned to the very laboratories that had produced the American bomb. Oppenheimer was an easy target.

New charges were now brought against Oppenheimer for his refusal to support the hydrogen bomb effort with sufficient enthusiasm. At first, he was too popular and influential to be tackled directly. His former students could be grilled, however. One, Bernard Peters, was named by Oppenheimer himself. In a closed hearing before HUAC, Oppenheimer said that Peters called himself a Communist fighting the Nazis. Oppenheimer concluded that Peters was still "dangerous." Peters denied the charges vigorously and wrote to his old teacher for clarification. Oppenheimer equivocated.

Close friends reproached Oppenheimer for testifying. Victor Weisskopf wrote to Oppenheimer in dismay: "[W]e are losing something that is irreparable. Namely confidence in *you* . . . whom so many regard as our representative in the best sense of the word."[29] Hans Bethe, Edward Condon, and even Edward Teller were horrified. Oppenheimer, confronted by Peters, apologized after a fashion. Writing to Weisskopf, Peters recalled: "He [Oppenheimer] said it was a terrible mistake. He was not prepared for any questions. He had never done anything as wrong." Peters felt "sad" to see Oppenheimer in such "moral despair."[30]

In the 1954 hearing, Oppenheimer himself was finally brought down by his pursuers. One, Lewis Strauss, was a wealthy businessman who had become a rear admiral during the war and then chairman of the Atomic Energy Commission (AEC). Strauss was touchy and vain. Oppenheimer unwisely mocked Strauss's views before Congress. The specific reason for the hearing was to inquire about Oppenheimer's refusal to support building the hydrogen bomb, though he was scarcely the only one to voice opposition — Bethe, Rabi, and James Bryant Conant, the president of Harvard, were also opposed. Oppenheimer's loyalty was not questioned, but he was nonetheless said to suffer a "susceptibility to influence." Rabi voiced support for Oppenheimer: We built you the A-bomb, "and what more do you want, mermaids?"[31] Edward Teller, on the other hand, testified that he did not feel "comfortable" with Oppenheimer's holding a security clearance. Teller was thereafter reviled by a large part of the physics community.

Oppenheimer, forced out of government service, went back to the Institute and physics, more serene in some ways, more troubled in others. He lived little more than a decade longer, dying of throat cancer in 1967. In his blunt way, Rabi insisted that the HUAC hearings were meant to kill Oppenheimer — and did. The Oppenheimer security trial is sometimes said to have inaugurated a new era in governmental suspicion and control of its scientists. Yet Oppenheimer must have known, during those uncomfortable interviews of 1943, that he had met his masters.

Oppenheimer succeeded in building the bomb, and was eventually disgraced. Heisenberg failed, and was received in triumph.

DANGEROUS KNOWLEDGE: THE NEW SECURITY ORDER

Einstein no sooner arrived in Princeton in 1933 than the FBI opened a file on him as a suspicious character. To keep him out of the country, the Woman Patriot group, made up of affluent right-wing women living in Washington, D.C., published a sixteen-page screed. They accused Einstein of being the true and "acknowledged world leader" of all Communist activity, outdoing "even Stalin himself" in this effort. Einstein meant to destroy all organized government, promote treason, organize unlawful "acts of rebellion against officers of the U.S. in time of war." He was also a charlatan; his relativity theory was nonsense; he was moreover an atheist.[32] This was the first item in Einstein's FBI file. Einstein read a copy of the publication and thought it so funny he answered it in print lightheartedly. The FBI did not find it amusing; its fantasies would be repeated a thousand times over in that file. The FBI added this warning when it forwarded the file to Army G-2 in 1940:

> In view of his radical background, this office would not recommend the employment of Dr. Einstein on matters of a secret nature, without a very careful investigation, as it seems unlikely that a man of his background could, in such a short time, become a loyal American citizen.[33]

That same year, Einstein became a citizen of the United States. He would contribute to the war effort minimally, as a consultant to the Navy, with his wild hair intact.

Even without Einstein's help, the Americans built the bomb and won the war. But an unexpected result was that a new and lasting conflict broke out around science and within it. Since they alone understood how to build such destructive weapons, physicists became indispensable to their governments as never before. The possession of such incalculably dangerous knowledge made them suspect — top security risks — to the authorities, who now could not do without them, but in many ways did not quite know what to do with them. For the scientists, there was an added sting of self-suspicion: After the bombs had burned away Japanese cities and their people, science itself became suspect to many of its creators, innocent no longer. Modern physics had earned a new and bitter pathos of its own. As early as 1945, Oppenheimer put it eloquently:

> We have made a thing, a most terrible weapon, that altered abruptly and profoundly the nature of the world. We have made a thing that by all the standards of the world we grew up in is an evil thing. And . . . we have raised again the question of whether science is good for men.[34]

Until the destruction of Hiroshima, the atom bomb was a project under the tightest wraps. The atom bomb was a weapon that would bring a truly "total" war and threaten the very existence of humanity anywhere on the globe. In the short run, the weapon gave the United States an advantage over rivals such Germany and the USSR.

By 1945, the United States had poured four billion dollars into the atom bomb project, an astonishing sum for that time. The project involved thousands of technicians, vast tracts in Tennessee and Washington, and Los Alamos in New Mexico. It was kept so secret that not even Vice President Harry S. Truman was told about it until he took the oath as president after Roosevelt died. The entire project was set up under control of the Army, with then Brigadier General Leslie Groves in charge. Army security did the obvious things: It surrounded Los Alamos with barbed wire, had sentries patrol the outskirts, set up censorship of mail. But all this was futile unless the physicists were themselves loyal and remained so. How much could these scientists be trusted? Neither security officials nor the scientists were prepared for the complexities involved.

The most valuable physicists were very much a foreign colony of savants. Fermi came from Italy; Wigner, von Neumann, and Teller from Hungary; Bethe from Germany, and Rudolf Peierls from Germany by way of England; Chadwick from England; Bohr from Denmark; Frisch from Austria by way of Denmark; Stanislaw Ulam from Poland; Vera Kistiakowsky from Russia. Their political views could be as puzzling to American security as their accents.

General Groves worried about having to deal with prima donnas, but except for Teller, there were few of those. Nor were there many troublesome political radicals. The Europeans had a closer experience with the extremist left and right and were apt to be politically conservative in the United States — Hungarians like Wigner or von Neumann, for example. Fermi left Italy because its anti-Semitism threatened his Jewish wife; in the United States, he did not bother much with politics. But two important scientists stood on the left. One was Klaus Fuchs, a German émigré to Britain, thereafter assigned to Los Alamos, with access to all secrets. The other was Robert Oppenheimer. The irony was that the Army and the FBI never suspected Fuchs of being a spy, though he passed secrets to the USSR from 1942 to 1949 and gave the Soviets all they needed to know about American know-how and progress; but they continually suspected and hounded Oppenheimer, who in fact did not pass any secrets. A further irony is that without Oppenheimer as director of Los Alamos, it is entirely possible that the atom bomb would never have been built, at least not so soon.

EPILOGUE

The Projects of Science

THE PATHOS OF SCIENCE LIES IN ITS DOUBLE NATURE: The scientist is at once free and strictly confined, individual but ultimately subsumed. This double role begins when the apprentice scientist starts the long and exacting effort to master the findings of that formidable (and always growing) army of predecessors. The energy of the young scientist, the intense interest, busy labor, and excitement of possible discovery naturally block off presentiments of eventually being an old lion in winter — and fortunately so, for the sake of science. Trying to make any sort of advance is strenuous enough without also contemplating being ultimately dislodged. Physics sees itself as a self-erasing discipline, concerned only with the leading edge of research.

Those no longer on the leading edge — whether a few years behind, or centuries — no longer have an independent existence, as, say, Shakespeare and Rembrandt continue to have. Einstein was at the leading edge until 1926, but thereafter became like those he himself had once helped supplant. One might say that scientists have two careers, the living one of progress and discovery, and the posthumous one — and in certain ways, the posthumous career can begin before death occurs.

Needless to say, science never advances very neatly. The time-lines of discovery move at very different speeds, and often in odd directions. While Einstein brought relativity to consummation in 1905, clarity about the atom progressed in fits and starts. The electron was discovered inside the atom in 1897; radioactive matter in 1896; the quantum in 1900. In 1911, Rutherford found the atomic nucleus; in 1913, Bohr showed that the stability of the atom required a quantum explanation. Quantum mechanics arrived in 1925. The physics of the nucleus began to catch up only in the 1930s.

One scientist can be trumped very quickly by a new advance — consider Schrödinger, who thought his wave equation of 1926 had rid physics of the plague of quantum mechanics, only to find he had been co-opted within a year. Newton was dead two centuries before his theory was supplanted. Some discoveries are tied not to an individual, but to a team, or are a mosaic of findings, supplanted bit by bit. Older scientists begin to harvest the limitations science set on them when they entered the fray, as if a custodianship.

Every advance costs an earlier achievement's demotion or displacement. Einstein revered Galileo, Newton, Maxwell, Lorentz, and Planck even as his findings dislodged each of them. He turned Newtonian gravitation into a special case of general relativity; used the Maxwellian field concept to supplant Newtonian mechanics; used Maxwell to radically revise older views of space and time, including Maxwell's own; routed the ether principle, which Lorentz clung to; transformed Planck's quantum concept from a "black body" concept into the vast new subject of quantum theory. (The reader can choose other verbs.) If he had succeeded as hoped with unification, the list would be much longer, beginning with a fundamental revision of what electromagnetism and quantum theory mean.

As the cutting edge advances, those once in the forefront of research are left behind. This can scarcely be lamented, since science would otherwise not keep advancing. Even a Newton or an Einstein will be dislodged. Homer, Bach, Botticelli — never. Science in this sense is one of the strangest human enterprises, imposing a limitation known nowhere else in thought or art.

None of this was lost on Einstein. His sharp and humorous sense of how he came to be regarded as a "petrified object" was part of his realism about what would happen to his — and everyone else's — place in science. If physics cannot be based on the field concept, he wrote to his old friend Besso in 1954, then "*nothing* remains of my entire castle in the air, gravitation theory included, [and of] the rest of modern physics."[1] Even if the field theory holds, it will be modified.

There are countless studies of genius and creativity, but the decline of great scientists is a largely uncharted subject. Some preliminary sifting is needed. If one asks why Einstein ceased being the Einstein who revolutionized physics, a few explanations have become familiar. First and expectedly, his gifts are said to have faded as he grew older. As this happened, the young man's strengths became the older man's handicaps. In the Swiss patent office and in Berlin, he was a loner, unusually stubborn, fiercely independent, self-isolated — all this was an instinctual wisdom about how to protect and fulfill his great gifts. But later, the stubbornness hardened into an obstinate clinging to fixed ideas; the self-isolation ignored new findings that could challenge his preconceptions; he became inflexible as his younger self never was. In this view, too, he was the victim of his own early success. General relativity — that single-handed triumph against all odds and cautious advice — made him overconfident that the same method could handle the new problem. His triumphant experience "seared" him, said Abraham Pais, a colleague of Einstein at Princeton. It kept him persisting, decade after decade, despite setbacks that should have warned him off.

Yet one may wonder. A quite different picture of the older Einstein also exists. Mathematical ability proverbially weakens early, but the aging Einstein kept his prowess. Peter G. Bergmann, one of his younger mathematical assistants in the 1930s, recalled:

> [What] impressed me — and remember that I was very young and Einstein in his fifties — was his tremendous creativeness . . . his sheer inventiveness of new approaches, of new mathematical tricks.[2]

Mathematics is not, however, the indispensable gift for a theoretical physicist — rather, physical intuition: the inner sense, judgment, hunch, perceptive inspiration, or however one names the inner gyroscopic instinct insisting that Nature supports this idea but not that one — and, of course, turns out to be right. In 1918, when (as noted earlier) Einstein rejected Hermann Weyl's unifying of gravity and electromagnetism as not corresponding to reality, Weyl made a telling remark: "The criticism," he replied to Einstein, "very much disturbs me, of course, since experience has shown that one can rely on your intuition."[3] Everyone thought the same, and rightly so: Everything Einstein had done since the age of twenty-six demonstrated it. It is why he could often proceed without benefit of laboratory experiments, using only his "thought experiments."

Did this supreme gift desert him in his later years? The American relativist John Wheeler of Princeton, who knew Einstein well from the 1930s, emphatically thinks not. In 1954, Wheeler hypothesized a "geon" — "a gravitating body made entirely of electromagnetic fields" — and sent his paper to Einstein for comment. Einstein, then seventy-five, thought about it awhile and said he doubted that a geon was stable; it took the much younger Wheeler several years to realize that Einstein was right.[4] Einstein's intuition, Wheeler said, was as "amazing" as always.

But even if all Einstein's powers did flag suddenly and disastrously, this explanation is still too limited. The outcome of any scientific career depends as much on what others accomplish: New discoveries can throw logs across the path, raise perplexities, find powerful new explanations. Einstein hugely admired Lorentz, and Lorentz was hardly bereft of his great powers in 1905, mainly because the young Einstein dismissed the ether to which Lorentz was so committed. Einstein revered Newton even more, but he wrote:

> Newton, forgive me: you found the only way which in your age was just about possible for a man with the highest powers of thought and creativity. The concepts which you created are guiding our thinking in physics even today, although we now know they will have to be replaced by others farther removed from the sphere of immediate experience, if we aim at a profounder understanding of relationships.[5]

What Newton did, what Einstein did after him, was possible only because science is a collective and cumulative enterprise.

The eighteenth-century poet Alexander Pope proclaimed: "God said, Let Newton be! and all was light." Newton preferred a more sober view. In 1675, he wrote to Hooke:

> You defer too much to my ability in searching into this subject. What Descartes did was a good step. You have added much. . . . If I have seen further it is by standing on the shoulders of giants.[6]

Science, said Robert Oppenheimer, is cumulative in "a quite special sense." Its findings are defined in terms of the objects and laws and ideas that were the science of its predecessors. What Galileo or Faraday did is, for working physicists, not history to be learned, but tools to be used: the physicist's very language, subject, definitions, rules, instruments, and foundations.

Einstein used these tools freely in search of his quest for unity. Still, he never believed that Nature can be made to yield her secrets given enough brute force (say, accelerators) and ingenuity. Einstein spoke instead, in a Goethean vein, of the "implacable smile of Nature," at once benign and mocking towards human efforts to fathom her mysteries. It may be that these efforts will always fall short, since Nature stands beyond. To those with eyes to see, the implacable smile foretells ultimate failure; any lifting of the veil is triumph enough. The smile, however, also bespeaks a benign posture, evidenced in what Einstein said was the ultimate mystery: that the universe is intelligible, and we can partake of that knowledge — and of that nobility. As he once said, "Nature conceals her secret through her essential nobility, not out of cunning."[7]

For Einstein, the defeat that science visits on its practitioners seemed as impersonal as the order of the universe. Most scientists probably agree, in principle. But the daily life of science is simply not geared to Einstein's exalted attitude. Intense competitiveness spurs the scientist to produce that extra surge of energy, labor, and intense concentration needed. It can spill over into unseemly scuffles for fame and prizes. Few work on the rarefied mountaintops of theory as Einstein did, commanding the grand view — the field is parceled instead into specialties and subspecialties, with teams of researchers instead of the lonely pioneer. By 1929, Dirac had made one extraordinary discovery after another; that year, he nonetheless wrote to Niels Bohr that quantum mechanics "will ultimately be replaced by something better, (and this applies to all physical theories)."[8]

The pathos of science does not always abide in the future. Often it doesn't wait, but invades the present as pressure, anxiety, doubt, or envy. Rivals may snatch away priority; someone else's research can devastatingly derail years of effort; or the supplanting future can appear in the here and now in the shape of an Einstein or Feynman, discouragingly quick, fertile, original; or invincibly sure-handed like Fermi or Rutherford. When the young Pauli and Heisenberg became Max Born's assistants at Göttingen, the older Born had such a presentiment. He said he couldn't match their genius.

If science is present-oriented, it also requires sustenance. It is vulnerable to the demands of its benefactors, whether the emergent German state on the cusp of World War I or the triumphant postnuclear United States.

Finally, the pathos can be shattering. Paul Ehrenfest, Einstein's beloved friend, committed suicide because he felt unable to keep up with the flood of new data and conundrums. His sense of being supplanted knotted unbearably and was — said Einstein — why he killed himself.

Einstein was also supplanted, but he was not shattered. In his old age, he became the ever more kindly, grandfatherly figure. He never considered his failing quest to be a tragedy. If many of his colleagues did, it was because he rejected quantum mechanics, the most vital new branch of physics. His last thirty years provoked steady and often unbridled opposition from cherished friends. Shortly before he died, Einstein wrote to Niels Bohr about banning nuclear weapons, but he couldn't resist teasing his old friend: "Don't frown like that," Einstein's letter began, "this is not about quantum mechanics." Einstein must have known that his personal quest for a unified theory had failed. In the end, belief trumped established science and yet left room for humor.

David Lindley, in his critical study *The End of Physics,* suggests that we have reached the end of what can be verified:

> What restrains the theorist from becoming wholly carried away by the attractions of some mathematical theory is the need to make predictions with it and to test them against the hard realities of the real world. But as, during this century, experiments in fundamental physics have become harder to do, more costly, and more consuming of time and manpower, this restraining empirical influence has been weakened.

Lindley believes that Einstein's general theory of relativity inaugurated a tendency to view "experimental verification [as] something of an afterthought."[9] Even its most fervent proponents acknowledge that superstring, the most encompassing of the string theories, is nearly impossible to test, even indirectly.[10] The same was said of early atomic theories; but by the time Russell visited Einstein and his friends in 1943, experiment was catching up to quantum physics. Einstein and Russell would spend the rest of their lives trying to wrest the atom from the industries of war.

What did the four great men speak of when they met at 112 Mercer Street? We will never know, of course. But speculation as to the topics is quite possible. They probably spoke little of the war: As Russell said, they were all in accord politically. Russell and Einstein disagreed over German war reparations, Einstein unable to forgive his former homeland. The doings of Los Alamos were sufficiently shrouded to have afforded little meat, although Pauli and Einstein were aware of the goings-on. Einstein and Pauli had just collaborated on a paper that touched on unified field theory and, despite their differences over quantum mechanics, found common ground — even suggesting, according to Pauli's biographer Charles Enz, an idea strikingly similar to the strings of string theory.[11] Gödel was writing his critique of Russell's mathematical logic for the *Living Philosophers* volume, but it is unlikely, given Gödel's taciturn nature and Russell's distance from his early work, that much was said of it.

Russell, however, may have listened attentively and probed deeply. His *Human Knowledge* was in the planning stages; its subject, the nature of scientific knowledge. In it, Russell would begin modestly and conclude even more modestly. "To discover the minimum principles required to justify scientific inferences is one of the main purposes of this book," he wrote in the Introduction. Having attempted to do so, he concludes, "[A]ll human knowledge is uncertain, inexact, and partial. To this doctrine we have not found any limitation whatever."

BIBLIOGRAPHY

Achinstein, Peter, and Stephen F. Barker, eds. *The Legacy of Logical Positivism: Studies in Philosophy of Science.* Baltimore: Johns Hopkins University Press, 1969.

Aczel, Amir D. *God's Equation: Einstein, Relativity and the Expanding Universe.* New York: Four Walls Eight Windows, 1999.

Aichelburg, Peter C., and Roman U. Sexl, eds. *Albert Einstein: His Influence on Physics, Philosophy and Politics.* Braunschweig, Wiesbaden: Vieweg, 1979.

Alonso, Marcelo, and Edward J. Finn. *Physics.* Reading, MA: Addison-Wesley, 1970.

Asimov, Isaac. *Understanding Physics.* 3 vols. New York: Barnes & Noble Books, 1993.

Atmanspacher, Harald, and Hans Prinas. "Pauli's Ideas on Mind and Matter in the Context of Contemporary Science." *The Journal of Consciousness Studies.* 13, 3, 5–50, 2006.

Ayer, A. J. *Russell and Moore: The Analytical Heritage.* Cambridge, MA: Harvard University Press, 1971.

Barnett, Lincoln. *The Universe and Dr. Einstein.* New York: New American Library, 1952.

Barrett, William, and Henry D. Aiken, eds. *Philosophy in the Twentieth*

Century, vol. 2, *The Rise of the British Tradition and Contemporary Analytic Philosophy.* New York: Random House, 1962.

Bergmann, Peter G. *The Riddle of Gravitation.* Revised and updated. Mineola, NY: Dover, 1992.

Berlin, Isaiah. *Concepts and Categories: Philosophical Essays.* New York: Viking, 1979.

Bernstein, Jeremy. *Albert Einstein and the Frontiers of Physics.* Oxford: Oxford University Press, 1996.

————, annotator. *Hitler's Uranium Club: The Secret Recordings at Farm Hall.* New York: Springer-Verlag, 2001.

Blanshard, Brand. *Reason and Analysis.* La Salle, IL: Open Court, 1964.

Bodanis, David. $E=mc^2$: *A Biography of the World's Most Famous Equation.* New York: Berkley, 2000.

Bondi, Hermann. *Relativity and Common Sense: A New Approach to Einstein.* Mineola, NY: Dover, 1980.

Born, Max, and Albert Einstein. *The Born-Einstein Letters.* Trans. Irene Born. New York: Macmillan, 2005.

Brian, Denis. *Einstein: A Life.* New York: John Wiley, 1996.

————. *The Unexpected Einstein: The Real Man Behind the Icon.* Hoboken, NJ: John Wiley, 2005.

Brod, Max. *The Redemption of Tycho Brahe.* Trans. Felix Warren Cross. New York: Alfred A. Knopf, 1928.

Bucky, Peter A. *The Private Albert Einstein.* Kansas City: Andrews and McMeel, 1992.

Bunch, Bryan. *Mathematical Fallacies and Paradoxes.* New York: Van Nostrand Reinhold, 1982.

Butcher, Sandra Ionno. "The Origins of the Russell-Einstein Manifesto." *Pugwash History Series,* no. 1, May 2005.

Bynum, W. F., E. J. Browne, and Roy Porter, eds. *Dictionary of the History of Science.* Princeton, NJ: Princeton University Press, 1984.

Calaprice, Alice. *The Einstein Almanac.* Baltimore: Johns Hopkins University Press, 2005.

Calder, Nigel. *Einstein's Universe: A Guide to the Theory of Relativity.* New York: Penguin, 1990.

Card, Charles R. "The Emergence of Archetypes in Present-Day Science and Its Significance for a Contemporary Philosophy of Nature." *Dynamical Psychology,* 1996.

Cassidy, David. *Einstein and Our World*. Amherst, NY: Prometheus Books, 2004.

———. *Uncertainty: The Life and Science of Werner Heisenberg*. New York: W. H. Freeman, 1992.

Casti, John L., and Werner DePauli. *Gödel: A Life of Logic*. Cambridge, MA: Perseus, 2000.

Chandrasekhar, S. *Truth and Beauty: Aesthetics and Motivations in Science*. Chicago: University of Chicago Press, 1987.

Chodorow, Joan. "Inner-Directed Movement in Analysis: Early Beginnings." *Inside Pages: The Jung Society of Seattle*, Spring 2005.

Clark, Ronald. *Einstein: The Life and Times*. New York: Harper & Row, 1984.

Concise Science Dictionary. 2nd ed. Oxford: Oxford University Press, 1991.

Cook, Alistair. *Six Men*. New York: Alfred A. Knopf, 1977.

Crawshay-Williams, Rupert. *Russell Remembered*. London: Oxford University Press, 1970.

Crease, Robert P., and Charles C. Mann. *The Second Creation: Makers of the Revolution in Twentieth-Century Physics*. New York: Macmillan, 1987.

Davies, Paul. *Superforce: The Search for a Grand Unified Theory of Nature*. New York: Simon & Schuster, 1984.

Dawson, John W., Jr., *Logical Dilemmas: The Life and Work of Kurt Gödel*. Wellesley, MA: A. K. Peters, 1997.

Dukas, Helen, and Banesh Hoffmann, eds. *Albert Einstein: The Human Side*. Princeton, NJ: Princeton University Press, 1986.

Eames, Elizabeth R. *Bertrand Russell's Dialogue with His Contemporaries*. Carbondale, IL: Southern Illinois University Press, 1989.

Edmonds, David, and John Eidinow. *Wittgenstein's Poker: The Story of a Ten-Minute Argument Between Two Great Philosophers*. New York: HarperCollins, 2001.

Einstein, Albert. *Autobiographical Notes*. Trans. and ed. Paul Arthur Schilpp. La Salle, IL: Open Court, 1992.

———. *The Collected Papers of Albert Einstein*. Multiple volumes. Ed. John J. Stachel et al. Princeton, NJ: Princeton University Press, 1987–.

———. "Concerning an Heuristic Point of View Toward the Emission and Transformation of Light." *Annalen der Physik*. 17, 132, 1905.

Translation into English, *American Journal of Physics,* vol. 33, no. 5, May 1965.

———. *Dear Professor Einstein: Albert Einstein's Letters to and from Children.* Ed. Alice Calaprice. Amherst, NY: Prometheus Books, 2002.

———. *Einstein on Peace.* Ed. Otto Nathan and Heinz Norden. New York: Simon & Schuster, 1960.

———. *Ideas and Opinions.* Trans. Sonja Bargmann. New York: Modern Library, 1994.

———. *Letters to Solovine.* Trans. Wade Baskin. New York: Carol Publishing, 1993.

———. *Out of My Later Years.* New York: Philosophical Library, 1950.

———. *Relativity: The Special and General Theory.* Trans. Robert W. Lawson. Mineola, NY: Dover, 2001.

———. *Sidelights on Relativity.* Trans. G.B. Jeffrey and W. Perrett. Mineola, NJ: Dover, 1983.

———. *The Theory of Relativity and Other Essays.* New York: MJF Books, 1950.

———. *The World as I See It.* Trans. Alan Harris. New York: Covici Friede, 1934.

———, and Michele Besso. *Correspondence, 1903–1955.* Paris: Hermann, 1972.

———, and Leopold Infeld. *The Evolution of Physics.* New York: Simon & Schuster, 1938.

Einstein, Maja. "Albert Einstein: A Biographical Sketch" (excerpt). *Resonance,* April 2000.

Enz, Charles P. *No Time to Be Brief: A Scientific Biography of Wolfgang Pauli.* Oxford: Oxford University Press, 2002.

———, and Karl von Meyenn, eds. *Wolfgang Pauli: Writings on Physics and Philosophy.* Trans. Robert Schlapp. Berlin: Springer-Verlag, 1994.

Epstein, Lewis Carroll. *Relativity Visualized.* San Francisco: Insight Press, 2000.

Feinberg, Barry, and Ronald Kasrils. *Bertrand Russell's America.* 2 vols. Boston: South End Press, 1983.

Feldman, Burton. *The Nobel Prize.* New York: Arcade, 2000.

Feuer, Lewis S. *Einstein and the Generations of Science.* 2nd ed. New Brunswick, NJ: Transaction, 1982.

Feynman, Richard P. *Six Easy Pieces.* New York: Perseus Books, 1983.

————. *Six Not-So-Easy Pieces: Einstein's Relativity, Symmetry, and Space-Time.* Reading, MA: Addison-Wesley, 1997.

Fölsing, Albrecht. *Albert Einstein.* Trans. Ewald Osers. New York: Penguin, 1997.

Frank, Philipp. *Einstein: His Life and Times.* Trans. George Rosen. New York: A. A. Knopf, 1947.

Franzen, Torkel. *Gödel's Theorem: An Incomplete Guide to Its Use and Abuse.* Wellesley, MA: A. K. Peters, 2005.

French, A. P., ed. *Einstein: A Centenary Volume.* Cambridge, MA: Harvard University Press, 1979.

————, and P.J. Kennedy, eds. *Niels Bohr: A Centenary Volume.* Cambridge, MA: Harvard University Press, 1985.

Friedman, Robert Marc. *The Politics of Excellence: Behind the Nobel Prize in Science.* New York: Henry Holt, 2001.

Fuchs, Walter R. *Physics for the Modern Mind.* Trans. Dr. M. Wilson and M. Wheaton. New York: Macmillan, 1967.

Gamow, George. *Thirty Years That Shook Physics: The Story of Quantum Theory.* Mineola, NY: Dover, 1985.

Gardner, Martin. *Relativity for the Million.* New York: Macmillan, 1962.

Goldstein, Rebecca. *Incompleteness: The Proof and Paradox of Kurt Gödel.* New York: Norton, 2006.

Goodchild, Peter. *J. Robert Oppenheimer: Shatter of Worlds.* New York: Fromm, 1985.

Grattan-Guinness, Ivor. *The Search for Mathematical Roots, 1870–1940: Logic, Set Theories and the Foundations of Mathematics from Cantor through Russell to Gödel.* Princeton: Princeton University Press, 2000.

————. *The Rainbow of Mathematics: A History of the Mathematical Sciences.* New York: W. W. Norton & Company, 2000.

Grayling, A.C. *Russell.* New York: Oxford University Press, 1996.

————. "Russell, Experience, and the Roots of Science." *The Cambridge Companion to Bertrand Russell.* Ed. Nicholas Griffin. Cambridge: Cambridge University Press, 2003.

Green, Brian. *The Elegant Universe.* London: Vintage, 2005.

Gribbin, John. *Companion to the Cosmos.* Boston: Little, Brown, 1996.

Griffin, Nicholas, and Alison R. Miculan, eds. *The Selected Letters of Bertrand Russell: The Public Years, 1914–1970.* London: Routledge, 2001.

Hampshire, Stuart N. *Modern Writers and Other Essays.* New York: Alfred A. Knopf, 1970.

Hawking, Stephen, ed. *God Created the Integers: The Mathematical Breakthroughs That Changed History.* Philadelphia: Running Press, 2005.

Heisenberg, Werner. *Physics and Beyond.* Trans. Arnold J. Pomerans. New York: Harper & Row, 1972.

Hilbert, David and Wilhelm Ackermann. *Principles of Mathematical Logic.* Trans. Lewis M. Hammond, George G. Leckie, and F. Steinhardt. New York: Chelsea, 1950.

Hoffmann, Banesh. *Albert Einstein, Creator and Rebel.* New York: Penguin, 1972.

————. *Relativity and Its Roots.* Mineola, NY: Dover, 1999.

Holton, Gerald. *Einstein, History, and Other Passions: The Rebellion Against Science at the End of the Twentieth Century.* Cambridge, MA: Harvard University Press, 2000.

————. *Thematic Origins of Scientific Thought: Kepler to Einstein.* Cambridge, MA: Harvard University Press, 1973.

Holton, Gerald, and Yehuda Elkana, eds. *Albert Einstein: Historical and Cultural Perspectives: The Centennial Symposium in Jerusalem.* Princeton, NJ: Princeton University Press, 1982.

Honner, John. *The Description of Nature: Niels Bohr and the Philosophy of Quantum Physics.* Oxford: Oxford University Press, 1987.

Howard, Don, and John Stachel, eds. *Einstein and the History of General Relativity.* Boston: Birkhauser, 1989.

————. *Einstein: The Formative Years, 1879–1909.* Boston: Birkhauser, 1998.

Hughes, Jeff. *The Manhattan Project: Big Science and the Atom Bomb.* Cambridge: Icon Books, 2002.

Huxley, Aldous. *Crome Yellow.* New York: Bantam, 1968.

Irvine, Andrew, ed. *Bertrand Russell: Critical Assessments of Leading Philosophers.* 4 vols. London: Routledge, 1999.

Jaki, Stanley. *A Mind's Matter.* Grand Rapids, MI: Wm. B. Eerdsmann, 2002.

Jerome, Fred. *The Einstein File: J. Edgar Hoover's Secret War Against the World's Most Famous Scientist.* New York: St. Martin's, 2002.

Johnson, Paul. *Modern Times.* New York: HarperCollins, 2001.

Junck, Robert. *Brighter than a Thousand Suns.* New York: Harcourt Brace Jovanovich, 1958.

Jung, Carl, and Wolfgang Pauli. *The Interpretation of Nature and the Psyche*. Trans. R.F.C. Hull and Pricilla Silz. London: Routledge & Kegan Paul, 1955.

Jungnickel, Christa, and Russell McCormmach. *Intellectual Mastery of Nature: Theoretical Physics from Ohm to Einstein*. 2 vols. Chicago: University of Chicago Press, 1986.

Kaku, Michio. *Einstein's Cosmos*. New York: W. W. Norton, 2004.

Klein, Etienne, and Marc Lachieze-Rey. *The Quest for Unity*. Trans. Axel Reisinger. Oxford: Oxford University Press, 1999.

Kragh, Helge. *Quantum Generations: A History of Physics in the Twentieth Century*. Princeton, NJ: Princeton University Press, 1999.

Kuhn, Thomas. *The Structures of Scientific Revolutions*. Chicago: University of Chicago Press, 1962.

Kuntz, Paul Grimley. *Bertrand Russell*. Boston: Twayne Publishers, 1986.

Lanczos, Cornelius. *Albert Einstein and the Cosmic World Order*. New York: John Wiley, 1965.

Laurikainen, K.V. *Beyond the Atom: The Philosophical Thought of Wolfgang Pauli*. Trans. Eugene Holman. Berlin: Springer-Verlag, 1988.

Levenson, Thomas. *Einstein in Berlin*. New York: Bantam, 2003.

Lieber, Lillian R. *The Einstein Theory of Relativity*. New York: Rinehart, 1945.

Lindley, David. *The End of Physics: The Myth of a Unified Theory*. New York: Basic Books, 1993.

Lindorff, David. *Pauli and Jung: The Meeting of Two Great Minds*. Wheaton, IL: Quest Books, 2004.

MacKinnon, Edward A., ed. *The Problem of Scientific Realism*. New York: Appleton Century Crofts, 1972.

McEvoy, J. P., and Oscar Zarate. *Introducing Quantum Theory*. Cambridge: Icon Books, 1999.

McLynn, Frank. "The Ghost in the Machine: Review of *The Spirit of Solitude*," *New Statesman & Society*, April 19, 1996, 9 (399) 36.

Meier, C. A., ed. *Atom and Archetype: The Pauli/Jung Letters, 1932–1958*. Princeton, NJ: Princeton University Press, 2001.

Michelmore, Peter. *The Swift Years: The Robert Oppenheimer Story*. New York: Dodd, Mead, 1969.

Millar, David, Ian John, and Margaret Millar. *The Cambridge Dictionary of Scientists*. Cambridge: Cambridge University Press, 1996.

Miller, Arther I. *Insights of Genius*. Cambridge, MA: MIT Press, 2000.

Mills, Robert. *Space, Time and Quanta: An Introduction to Contemporary Physics.* New York: W. H. Freeman, 1994.

Monk, Ray. *Bertrand Russell: The Ghost of Madness, 1921–1970.* New York: Free Press, 2001.

———. *Bertrand Russell: The Spirit of Solitude, 1872–1921.* New York: Free Press, 1996.

———. "Cambridge Philosophers: Russell." *Royal Institute of Philosophy Online.* http:www.royalinstitutephilosophy.org/articles/article.php?id=3 (accessed April 6, 2007).

Moorehead, Caroline. *Bertrand Russell: A Life.* New York: Viking, 1992.

Morrell, Ottoline. *Memoirs: A Study in Friendship 1873–1915.* Ed. Robert Gathome-Hardy. New York: Alfred A. Knopf, 1964.

Nagel, Ernest, and James R. Newman. *Gödel's Proof.* New York: New York University Press, 1973.

Nasar, Sylvia. *A Beautiful Mind.* New York: Simon & Schuster, 1998.

Oppenheimer, Robert. *Letters and Recollections.* Ed. Alice Kimball Smith and Charles Weiner. Stanford, CA: Stanford University Press, 1995.

Oxford Dictionary of Scientists. Oxford: Oxford University Press, 1999.

Pais, Abraham. *Inward Bound of Matter and Forces in the Physical World.* Oxford: Oxford University Press, 1986.

———. *Niels Bohr's Times, in Physics, Philosophy, and Polity.* Oxford: Oxford University Press, 1991.

———. *Subtle Is the Lord.* Oxford: Oxford University Press, 1982.

Parker, Barry. *Einstein's Dream: The Search for a Unified Theory of the Universe.* New York: Plenum, 1988.

Pauli Archives: CERN. http:cdsweb.cern.ch/collection/Pauli%20Archives.

Pears, D.F. *Bertrand Russell and the British Tradition in Philosophy.* New York: Random House, 1967.

Peat, F. David. *From Certainty to Uncertainty: The Story of Science and Ideas in the Twentieth Century.* Washington, DC: Joseph Henry Press, 2002.

Penrose, Roger. *The Emperor's New Mind: Concerning Computers, Minds, and the Laws of Physics.* New York: Penguin, 1991.

Perricone, Mike. "How to Make a Neutrino Beam." *Fermi News,* November 19, 1999, 22 (22) 1.

Popkin, Richard H., ed. *Columbia History of Western Philosophy.* New York: MJF Books, 1999.

Popović, Milan. *In Albert's Shadow: The Life and Letters of Mileva Marić,*

Einstein's First Wife. Baltimore, MD: Johns Hopkins University Press, 2003.

Potter, Michael. *Reason's Nearest Kin: Philosophies of Arithmetic from Kant to Carnap.* Oxford: Oxford University Press, 2000.

Powers, Thomas. *Heisenberg's War.* Cambridge, MA: Da Capo Press, 2000.

Regis, Ed. *Who Got Einstein's Office? Eccentricity and Genius at the Institute for Advanced Study.* Reading, MA: Addison-Wesley, 1987.

Rhodes, Richard. *The Making of the Atom Bomb.* New York: Simon & Schuster, 1988.

Robertson, Robin. *C. G. Jung and the Archetypes of the Collective Unconscious.* New York: Peter Lang, 1987.

Rose, Paul Lawrence. *Heisenberg and the Nazi Atomic Bomb Project: A Study in German Culture.* Berkeley: University of California Press, 1998.

Rosen, Stanley. *The Limits of Analysis.* New Haven, CT: Yale University Press, 1980.

Rothman, Tony. *Instant Physics: From Aristotle to Einstein, and Beyond.* New York: Fawcett, 1995.

Royal, Denise. *The Story of J. Robert Oppenheimer.* New York: St. Martin's, 1969.

Rucker, Rudolf v. B. *Geometry, Relativity and the Fourth Dimension.* Mineola, NY: Dover, 1977.

Russell, Bertrand. *The ABC of Relativity.* New York: New American Library, 1959.

————. *The Autobiography of Bertrand Russell.* London: Routledge, 1998.

————. *The Basic Writings of Bertrand Russell, 1903–1959.* Ed. Robert E. Egner and Lester E. Denonn. New York: Simon & Schuster, 1961.

————. *The Collected Papers of Bertrand Russell,* vol. 11, *Last Philosophical Testament.* Ed. John G. Slatter. London: Routledge, 1997.

————. *Essays on Language, Mind and Matter, 1919–26.* Ed. John Passmore. Vol. 9 of *The Collected Papers of Bertrand Russell.* London: Routledge, 1988.

————. *A History of Western Philosophy.* New York: Simon & Schuster, 1972.

————. *Human Knowledge: Its Scope and Limits.* London: Routledge, 2003.

——. *Introduction to Mathematical Philosophy.* Mineola, NY: Dover, 1993.

——. *My Philosophical Development.* London: Allen and Unwin, 1959.

——. *Our Knowledge of the External World.* New York: New American Library, 1960.

——. *Portraits from Memory and Other Essays.* New York: Simon & Schuster, 1956.

——. *The Principles of Mathematics.* New York: W. W. Norton, 1996.

——. *The Problems of Philosophy.* Mineola, NY: Dover, 1999.

——. *Selected Letters.* Ed. Nicholas Griffin. 2 vols. London: Allen Lane, 1992–2001.

Ryan, Alan. *Bertrand Russell: A Political Life.* New York: Farrar, Straus and Giroux, 1988.

Sayen, Jamie. *Einstein in America: The Scientist's Conscience in the Age of Hitler and Hiroshima.* New York: Crown, 1985.

Schilpp, Paul A., ed. *Albert Einstein: Philosopher-Scientist.* 2nd ed. New York: Tudor, 1951.

——. *The Philosophy of Bertrand Russell.* 3rd ed. New York: Tudor, 1951.

Schirmacher, Wolfgang, ed. *German Essays on Science in the Twentieth Century.* New York: Continuum, 1996.

Schweber, S.S. *In the Shadow of the Bomb.* Princeton, NJ: Princeton University Press, 2000.

——. Review of Jagdish Mehra and Helmut Rechenberg, *The Historical Development of Quantum Theory,* vol. 6, *The Completion of Quantum Mechanics, 1926–1941,* in *Physics Today,* November 2001. http:www.physicists.org/.

Schwerin, Alan. *Bertrand Russell on Nuclear War, Peace, and Language: Critical and Historical Essays.* Westport, CT: Praeger, 2002.

Shanker, Stuart G., ed. *Philosophy of Science, Logic and Mathematics in the Twentieth Century,* vol. IX of *Routledge History of Philosophy.* London: Routledge, 1996.

Shapiro, Stewart, ed. *The Oxford Handbook of Philosophy of Mathematics and Logic.* Oxford: Oxford University Press, 2005.

Shoenman, Ralph, ed. *Bertrand Russell: Philosopher of the Century.* Boston: Little, Brown, Atlantic Monthly Press Book, 1967.

Singh, Jagjit. *Great Ideas of Modern Mathematics: Their Nature and Use.* Mineola, NY: Dover, 1959.

Snow, C.P. *The Physicists: A Generation That Changed the World.* London: Macmillan, 1981.

Sonnert, Gerhard. *Einstein and Culture.* Amherst, NY: Prometheus, 2005.

Steiner, George. *Extra-Territorial: Papers on Literature and the Language Revolution.* New York: Atheneum, 1971.

Stevens, Graham. *The Russellian Origins of Analytical Philosophy: Bertrand Russell and the Unity of the Proposition.* London: Routledge, 2005.

Tait, Katharine. *My Father, Bertrand Russell.* New York: Harcourt Brace Jovanovich, 1975.

Tonnelaf, Marie-Antoinette. *The Principles of Electromagnetic Theory and of Relativity.* Trans. Arthur J. Knodel. New York: Gordon and Breach, 1966.

Toulmin, Stephen. *Return to Reason.* Cambridge, MA: Harvard University Press, 2001.

Tucci, Niccolo. "The Great Foreigner." *The New Yorker,* November 22, 1947, 43–44.

Ulam, Stanislaw. *Adventures of a Mathematician.* Berkeley: University of California Press, 1991.

von Baeyer, Hans Christian. *The Fermi Solution: Essays on Science.* New York: Random House, 1993.

von Meyenn, Karl, and Engelbert Schucking. "Wolfgang Pauli." *Physics Today,* February 2001. http://www.physicstoday.org/ (accessed April 6, 2007).

Walker, Mark. *Nazi Science: Myth, Truth, and the German Atomic Bomb.* New York: Plenum Press, 1995.

Wang, Hao. *Reflections on Kurt Gödel.* Cambridge, MA: MIT Press, 1995.

Weidlich, Thom. *Appointment Denied: The Inquisition of Bertrand Russell.* Amherst, NY: Prometheus Books, 2000.

Weinberg, Steven. "The Search for Unity: Notes for a History of Quantum Field Theory." *Daedalus: Journal of the American Academy of Arts and Sciences,* Fall 1977, vol. II, 17–35.

Weisskopf, Victor. *The Joy of Insight: Passions of a Physicist.* New York: HarperCollins, 1991.

Wheeler, John A. *Geons, Black Holes, and Quantum Foam: A Life in Physics.* New York: Norton, 1998.

Whitaker, Andrew. *Einstein, Bohr, and the Quantum Dilemma.* Cambridge: Cambridge University Press, 1996.

Whitrow, G. J., ed. *Einstein: The Man and His Achievement.* Mineola, NY: Dover, 1967.

Wigner, Eugene. "The Unreasonable Effectiveness of Mathematics in the Natural Sciences." *Communications in Pure and Applied Mathematics,* vol. 13, no. 1, February 1960.

Wilson, Jane S., and Charlotte Serber. *Standing By and Making Do.* Los Alamos, NM: Los Alamos Historical Society, 1988.

Wistrich, Robert S. *Who's Who in Nazi Germany.* London: Routledge, 1995.

Wood, Alan. "Russell's Philosophy: A Study of Its Development," in *Bertrand Russell: Critical Assessments of Leading Philosophers,* vol. 1. Ed. Andrew Irvine. London: Routledge, 1999.

Yourgrau, Palle. *Gödel Meets Einstein: Time Travel in the Gödel Universe.* Chicago: Open Court, 1999.

———. *A World Without Time: The Forgotten Legacy of Gödel and Einstein.* New York: Basic Books, 2005.

NOTES

INTRODUCTION

1. Hao Wang, *Reflections on Kurt Gödel* (Cambridge, MA: MIT Press, 1995), 112.

PART 1

1. Bertrand Russell, *The Autobiography of Bertrand Russell* (London: Routledge, 1998), 466.

2. Ronald Clark, *Einstein: The Life and Times* (New York: HarperCollins, 1984), 642.

3. *Autobiography*, 155.

4. See Satoshi Kanazawa, "Why Productivity Fades with Age: The Crime-Genius Connection," *Journal of Research in Personality* (2003), 37: 257–72. See also Dean Keith Simonton, *Scientific Genius: A Psychology of Science* (Cambridge: Cambridge University Press, 1988).

5. Albrecht Fölsing, *Albert Einstein: A Biography*, trans. Ewald Osers (New York: Viking, 1997), 694.

6. S. Chandrasekhar, *Truth and Beauty: Aesthetics and Motivations in Science* (Chicago: University of Chicago Press, 1987), 48.

7. In *The Structures of Scientific Revolutions* (Chicago: University of Chicago Press, 1962), Kuhn argues against a conventional notion of progress in science as if each successive scientific theory were simply "a better representation of what nature is really like" (206). He takes a more relativistic stance, and his "paradigm shifts" are more complexly layered.

Thus, he "do[es] not doubt, for example, that Newton's mechanics improves on Aristotle's and that Einstein's improves upon Newton's as instruments for puzzle-solving. But I can see in their succession no coherent direction of ontological development. On the contrary, in some important aspects, but by no means in all, Einstein's general theory of relativity is closer to Aristotle's than either of them is to Newton's" (206–7). Yet each developed his theory within the historical context of a collectively "created" paradigm.

8. Quoted in Banesh Hoffmann, *Albert Einstein, Creator and Rebel* (New York: Penguin, 1972), 257.

9. Bertrand Russell, *My Philosophical Development* (London and NY: Allen and Unwin, 1959), 57.

10. Quoted in Charles P. Enz, *No Time to Be Brief: A Scientific Biography of Wolfgang Pauli* (Oxford: Oxford University Press, 2002), 355.

11. Ibid., 392.

12. Folsing, 679. Translation revised by Burton Feldman.

13. Ibid., 648.

14. Ibid., 688.

15. Ibid., 690.

PART 2

1. Denis Brian, *Einstein: A Life* (New York: John Wiley & Sons, 1996), 276.

2. Albrecht Fölsing, *Albert Einstein,* trans. Ewald Osers (New York: Penguin, 1997), 651, 330.

3. Albert Einstein and Michele Besso, *Correspondence, 1903–1955* (Paris: Hermann, 1972), 538. Translation by Burton Feldman.

4. Niccolo Tucci, "The Great Foreigner," in *The New Yorker,* November 22, 1947.

5. Maja Einstein, "Albert Einstein: A Biographical Sketch (Excerpt)," *Resonance,* April 2000, 113.

6. Ibid., 115.

7. Fölsing, 23.

8. Ibid., 17.

9. Ibid., 56.

10. Thomas Levenson, *Einstein in Berlin* (New York: Bantam, 2003), 12.

11. *Albert Einstein, The Human Side,* ed. Helen Dukas and Banesh Hoffmann (Princeton: Princeton University Press, 1986), 54.

12. Fölsing, 33.

13. Fölsing, 114–115.

14. Gerald Holton, *Einstein, History, and Other Passions* (Woodbury, NY: AIP Press, 1995), 62.

15. Fölsing, 334.

16. Max Brod, *The Redemption of Tycho Brahe,* trans. Felix Warren Crosse (New York: Alfred A. Knopf, 1928), 89–90.

17. Ibid., 154.

18. Fölsing, 283.

19. *Dear Professor Einstein: Albert Einstein's Letters to and from Children,* ed. Alice Calaprice (Amherst, NY: Prometheus Books, 2002), 140.

20. Albert Einstein, *Ideas and Opinions,* trans. Sonja Bargmann (New York: Modern Library, 1994), 108.

21. Fölsing, 349.

22. Abraham Pais, *Subtle Is the Lord,* (Oxford: Oxford University Press, 1982), 308.

23. Albert Einstein, *Collected Papers: The Berlin Years, Correspondence,* vol. 8, part A, ed. R. Schulmann (Princeton: Princeton University Press, 1997), xxxvii.

24. Philipp Frank, *Einstein: His Life and Times* (New York: A.A. Knopf, 1947), 106.

25. Fölsing, 343.

26. Ibid., 344–45. Fulda was a Jew who committed suicide in 1939. He wrote several plays that were adapted to the screen, including *Two-Faced Woman,* a poorly received comedy that was to be Greta Garbo's last film.

27. Fölsing, 345.

28. *Einstein on Peace,* ed. Otto Nathan and Heinz Norden (New York: Simon & Schuster, 1960), 12.

29. *The Collected Papers,* vol. 8, part A, 188.

30. Pais, 313, notes that sometime during the war the Berlin military chief of staff sent a list of pacifists, including Einstein, to the police.

31. Einstein, *Collected Papers,* vol. 8, 210.

32. Ibid., 342.

33. Otto Nathan and Heinz Norden, *Einstein on Peace* (New York: Simon & Schuster, 1960), 8.

34. For a frank discussion of Einstein's wartime activities, see Fölsing, 398ff.

35. Ibid., 398–99.

36. Pais, *Subtle,* 307.

37. Fölsing, 458.

38. *The Born-Einstein Letters,* trans. Irene Born (New York: Macmillan, 2005; first published 1971), 12.

39. *The Born-Einstein Letters,* 4.

40. "Our Debt to Zionism," in *Out of My Later Years,* 262.

41. Fölsing, 494.

42. Ibid., 495.

43. Ibid., 497.

44. Ibid., 515.

45. Ibid., 519.

46. Ibid., 464, 520.

47. The world could seem very small in those days: Samuel was the Home Secretary — head of police and security — who had hounded Russell for his peace work and declared that "There is no question, of course, that he is an enemy agent" during World War I. See Ray Monk, *Bertrand Russell: The Spirit of Solitude, 1872–1921* (New York: Free Press, 1996), 474. Samuel had Russell fined and later imprisoned. Russell's older

brother Frank was then the second Earl Russell; Samuel had been Frank Russell's "fag" — student servant — when both attended Winchester.

48. Fölsing, 594–95.

49. Gerald Holton and Yehuda Elkana, eds., *Albert Einstein: Historical and Cultural Perspectives, The Centennial Symposium in Jerusalem* (Princeton, NJ: Princeton University Press, 1982), 294.

50. Fölsing, 733.

51. See David Cassidy, *Einstein and Our World* (New York: Humanity Books, 2004), 82–85, for insight into how quantum mechanics blossomed in the ruins of postwar Germany.

52. Ronald William Clark, *Einstein: The Life and Times* (New York: Avon Books, 1984), 494–95.

53. Fölsing, 661.

54. *Born-Einstein Letters,* 112.

55. Fölsing, 679; Pais, 452.

56. See Sylvia Nasar, *A Beautiful Mind* (New York: Simon & Schuster, 1998), 49ff. for a richly detailed description of Princeton in the late 1940s. The Oral History Project of the Princeton Mathematics Community of the 1930s includes a brief history of Fine Hall (now called Jones Hall) at http://infoshare1.princeton.edu/libraries/firestone/rbsc/finding_aids/mathoral/pm06.htm.

57. Sandra Ionno Butcher, "The Origins of the Russell-Einstein Manifesto," *Pugwash History Series* no. 1, May 2005, 14.

58. Caroline Morehead, *Bertrand Russell: A Life* (New York: Viking, 1992), 204–6.

59. Russell, *Autobiography,* 445–47.

60. Ibid., 442.

61. Fölsing, 25; Ray Monk, *Bertrand Russell: The Spirit of Solitude 1872–1921* (New York: Free Press, 1996), 49.

62. Fölsing, 99.

63. Einstein, *Ideas and Opinions,* 10.

64. Thomas Levenson, *Einstein in Berlin* (New York: Random House, 2004), 9.

65. Moorehead, 238.

66. *Autobiography,* 334.

67. Moorehead, 238.

68. Aldous Huxley, *Crome Yellow* (New York: Bantam, 1968), 14.

69. Einstein, *Ideas and Opinions,* 82.

70. Ibid., 83.

71. Denis Brian, *The Unexpected Einstein: The Real Man Behind the Icon* (Hoboken, NJ: John Wiley & Sons, 2005), 24.

72. Russell, *Autobiography,* 216–17.

73. Ibid., 9.

74. Einstein, *Ideas and Opinions,* 320.

75. Rupert Crawshay-Williams, *Russell Remembered* (London: Oxford University Press, 1970), 18.

76. *Autobiography,* 17.

77. Moorehead, 19.

78. *Autobiography,* 29, 41.

79. Bertrand Russell, "My Mental Development," in *The Philosphy of Bertrand Russell,* ed. Paul A. Schilpp, 3rd ed. (New York: Tudor Publishing, 1951), 41.

80. His grandmother urged Russell to try "detaching your mind from the one subject [Alys] and bidding it range over others. . . ." *The Selected Letters of Bertrand Russell: Volume 1, The Private Years, 1884–1914,* ed. Nicholas Griffin (London: Penguin, 1992), 528. Lady Russell was exceedingly shrewd and manipulative, though subject to delusions.

81. Ray Monk, *Bertrand Russell: The Spirit of Solitude* (New York: Free Press, 1996), 196.

82. *Autobiography,* 303.

83. Michael Foot, introduction to *Autobiography,* x.

84. Monk, *Spirit,* 256.

85. Ibid., 257.

86. *Autobiography,* 82.

87. Ibid., 256.

88. Ibid.

89. Monk's highly critical two-volume biography of Russell elicited angry retorts. Michael Foot called it an "assault on Bertrand Russell's reputation" full of "malevolence." (Introduction, *Autobiography,* ix–x.)

90. Crawshay-Williams, 157.

91. Monk, *Spirit,* 135.

92. Caroline Moorehead, *Bertrand Russell: A Life* (New York: Viking, 1992), 172.

93. Monk, *Spirit,* 297, 300.

94. Ibid., 175.

95. Monk, *Spirit,* 295.

96. *Autobiography,* 329.

97. Ibid., 246.

98. Ottoline Morrell, *Memoirs: A Study in Friendship 1873–1915,* ed. Robert Gathome-Hardy (New York: Alfred A. Knopf, 1964), 276.

99. *Autobiography,* 243, 245.

100. Ibid., 245.

101. Ibid., 244.

102. Ray Monk, *Bertrand Russell: The Ghost of Madness, 1921–1970* (New York: The Free Press, 2000), 95.

103. Moorehead, 446.

104. *The Selected Letters of Bertrand Russell: The Public Years, 1914–1970,* vol. 2, ed. Nicholas Griffin (New York: Routledge, 2001), 353.

105. Monk, *Spirit,* 142.

106. Ibid., 147.

107. Ibid., 183.

108. *Autobiography,* 167–68.

109. Ibid., 210.

110. Roger Kimball, "Love, Logic & Unbearable Pity: The Private Bertrand Russell," *New Criterion Online,* http://www.newcriterion.com/archive/11/sept92/brussell.htm.

111. Russell included an addendum on Gödel's essay when the collection was reprinted in 1965.

112. Russell, *Autobiography,* 238.

113. Monk, *Spirit,* 126.

114. Ibid., 382.

115. Moorehead, 213.

116. Russell, *Autobiography,* 277.

117. Moorehead, 243.

118. A scheme by which some thirty-four objectors were to be sent to the front, where their refusal to take up arms would be deemed "desertion," punishable by death, was finally averted through the efforts of Russell and his colleagues. See Alan Ryan, *Bertrand Russell: A Political Life* (New York: Farrar, Straus and Giroux, Hill and Wang, 1988), 56.

119. Monk, *Spirit,* 457.

120. Ibid., 459.

121. Ibid., 456.

122. Moorehead, 213.

123. Ibid., 254.

124. Monk, *Spirit,* 466.

125. Ibid., 474.

126. Ibid., 471.

127. Ibid., 521–23.

128. *Autobiography,* 258.

129. Monk, *Spirit,* 532–34.

130. Ibid., 466.

131. *Autobiography,* 9, 727–28.

132. Frank McLynn, "The Ghost in the Machine: Review of *The Spirit of Solitude,*" *New Statesman & Society,* April 19, 1996, 9 (399) 36.

133. Stuart Hampshire, *Modern Writers and Other Essays* (New York: Alfred A. Knopf, Borzoi Books, 1970), 115; Michael Foot's introduction to Russell's *Autobiography,* ix.

134. Alan Wood, "Russell's Philosophy: A Study of Its Development," in *Bertrand Russell: Critical Assessments of Leading Philosophers,* vol. 1, ed. Andrew Irvine (London: Routledge, 1999), 86.

135. Russell, Preface to *Human Knowledge* (London: Routledge, 2003), 5.

136. Moorehead, 432.

137. *Collected Papers,* vol. 11, 114.

138. Ray Monk, "Cambridge Philosophers: Russell," *Royal Institute of Philosophy,* online http://www.royalinstitutephilosophy.org/articles/article.php?id=3. Accessed April 6, 2007.

139. "Russell, Experience, and the Roots of Science," in *The Cambridge Companion to Bertrand Russell,* ed. Nicholas Griffin (Cambridge: Cambridge University Press, 2003), 449–50.

140. John W. Dawson, *Logical Dilemmas: The Life and Work of Kurt Gödel* (Wellesley, MA: A. K. Peters, 1997), 204n.

141. See Dawson, 188. Gödel is quoted as saying *furchtbar herzig.* "Terribly cute" is Burton Feldman's translation.

142. Hao Wang, *Reflections on Kurt Gödel* (Cambridge, MA: MIT Press, 1995), 31.

143. *Albert Einstein: Historical and Cultural Perspectives, The Centennial Symposium in Jerusalem,* ed. Gerald Holton and Yehuda Elkana (Mineola, NY: Dover, 1982), 422.

144. Palle Yourgrau, *A World without Time: The Forgotten Legacy of Gödel and Einstein* (New York: Basic Books, 2005), 4.

145. Stanislaw Ulam, *Adventures of a Mathematician* (Berkeley: University of California Press, 1991), 76.

146. Dawson, 111, 234.

147. Ibid., 31.

148. Ibid., 34, 187.

149. Ibid., 130.

150. Ibid., 153, 187.

151. Ibid., 158.

152. Wang, 6.

153. Dawson, 140–41.

154. Ibid., 142.

155. Ibid., 142.

156. Ibid., 91.

157. Ibid., 135.

158. Ibid., 262.

159. Ibid., 251.

160. Ibid., 201.

161. Ibid., 98. See also Dawson's notion of the "paradox of paranoia," 265–66.

162. Wang, 214, quoting a letter from Gödel to his mother dated July 23, 1961.

163. Yourgrau's previous works on Gödel and Einstein were written for specialists, not, as he notes in *A World Without Time,* for "normal readers" (p. vii).

164. The distant ancestor of Gödel's effort here is perhaps Socrates, whose insistence that he did not know was an astonishingly fertile innovation, for it results in a dialectic (recursive) movement which sees any claim to know as the basis for a new test of itself. Although Gödel claimed to be a Platonist, at least a mathematical one, his work suggests the "incompletability" of thought that Socrates taught by his use of irony and myth.

165. "Ich war so dumm wenn [als] ich jung war!" Quoted in and translated by Charles Enz, *No Time to Be Brief* (Oxford: Oxford University Press, 2002), 117. Enz is the preeminent authority on Pauli's life, and this section is greatly indebted to his work. The discovery of electron spin in 1925 is told, delightfully, from another point of view in Samuel Goudsmit's 1971 lecture found at http://www.ilorentz.org/history/spin/

goudsmit.html. The codiscoverers were S. A. Goudsmit and G. E. Uhlenbeck.

166. Ibid., 491–92.

167. Werner Heisenberg, IAEA Bulletin Special Supplement (1968), 45, quoted in Karl von Meyenn and Engelbert Schucking, "Wolfgang Pauli," *Physics Today* (February 2001) http://www.physicstoday.org/ (accessed April 6, 2007).

168. Pauli Archive, CERN, http://documents.cern.ch//archive/electronic/other/pauli_vol1//born_0027.pdf, trans. Burton Feldman. Also quoted and translated in Enz, 394.

169. Enz points out that Sigmund Freud spent years as a medical doctor at a Viennese hospital before, in 1902, he was able to network and publish his way into a professorship at the University of Vienna (Enz, 8).

170. Ibid., 4.

171. Bertrand Russell, *An Outline of Philosophy* (London: Routledge, 1993), 235.

172. von Meyenn, 2001.

173. Enz, 49.

174. K. V. Laurikainen, *Beyond the Atom: The Philosophical Thought of Wolfgang Pauli* (Berlin: Springer-Verlag, 1988), 4.

175. Richard Courant, quoted in Enz, 87.

176. Ibid., 88.

177. Silvan Schweber, review of Jagdish Mehra and Helmut Rechenberg, *The Historical Development of Quantum Theory, Volume 6: The Completion of Quantum Mechanics, 1926–1941,* in *Physics Today* (November 2001), http://www.physicists.org/.

178. Enz, 92.

179. Ibid., 107.

180. Enz recounts in great detail the "curious history of spin." In later years, a battle of words erupted between Pauli and Goudsmit over whether a note published by Pauli in 1924 should have been credited to Goudsmit and Uhlenbeck. In the note, Enz argues, Pauli suggests the idea of *nuclear* spin. See Enz, especially 116–19.

181. Werner Heisenberg, *Physics and Beyond* (New York: Harper & Row, 1971), 61.

182. Enz, 129.

183. Karl von Meyenn and Engelbert Schucking allude to rumors that Pauli's letters "were taken from Heisenberg when he was arrested by the British in 1945" and thus may have survived the war. See von Meyenn.

184. Enz, 215.

185. Mike Perricone, "How to Make a Neutrino Beam," *Fermi News,* November 19, 1999, 22 (22) 1. Available online at http://www.fnal.gov/pub/ferminews/Ferminews99-11-19.pdf.

186. Enz, 195.

187. See Joan Chodorow, "Inner-Directed Movement in Analysis: Early Beginnings," *Inside Pages: The Jung Society of Seattle,* Spring 2005, 15.

188. Enz, 210.

189. Ibid., 224.

190. Ibid., 243.

191. Pauli was not Jung's only inspiration from the world of physics. Jung and Einstein met during the latter's stay in Zurich from 1909–13. In a letter, Jung recalled, "it was he [Einstein] who first started me off thinking about a possible relativity of time as well as space, and their psychic conditionality. More than thirty years later, this stimulus led to my relation with the physicist Professor W. Pauli and to my thesis of psychic synchronicity." Quoted in Charles R. Card, "The Emergence of Archetypes in Present-Day Science and Its Significance for a Contemporary Philosophy of Nature," *Dynamical Psychology,* 1996, available online: http://www.goertzel.org/dynapsyc/index.htm#1996.

192. From Enz's conversation with Franca Pauli in 1971. See Enz, 286.

193. Pais, 347.

PART 3

1. Brian Greene paraphrases Ernst Rutherford's admonition "if you can't explain a result in simple, nontechnical terms, then you don't really understand it." Not that it isn't true, Greene hastens to add (he is writing in most laudatory terms of string theory). A theory's truth and our true un-

derstanding of it are separate things. *The Elegant Universe* (New York: Vintage, 2005), 203.

2. Albert Einstein, *Sidelights on Relativity*, trans. G. B. Jeffrey and W. Perrett (Mineola, NY: Dover, 1983), 15.

3. "The Unreasonable Effectiveness of Mathematics in the Natural Sciences," in *Communications in Pure and Applied Mathematics*, vol. 13, no. 1 (February 1960), 7.

4. Albrecht Fölsing, *Albert Einstein*, trans. Ewald Osers (New York: Penguin, 1997), 390.

5. Ray Monk, *Bertrand Russell: The Ghost of Madness, 1921–1970* (New York: Free Press, 2001), 269.

6. Quoted in Torkel Franzen, *Gödel's Theorem: An Incomplete Guide to Its Use and Abuse* (Wellesley, MA: A. K. Peters, 2005), 112–13.

7. Ibid., 113.

8. For detailed discussions of the proofs, the nonmathematical reader is referred elsewhere. Several able mathematicians have rendered Gödel as accessible as he can be to the nonmathematician. First and foremost are Ernest Nagel and James R. Newman, whose *Gödel's Proof* (dedicated to Bertrand Russell!), written five years before Gödel's death, first made Gödel possible for those with limited (though still hearty) mathematics. Since then, Gödel and incompleteness have entered the lay world via the bestselling *Gödel, Escher, Bach: An Eternal Golden Braid* by Richard Hofstadter. A recent twist is Palle Yourgrau's *A World Without Time*, pairing Gödel and Einstein. Hofstadter is a cognitive scientist and Yourgrau a philosopher. Mathematicians continue to proffer "accessible" translations of the theorems. Two recent forays — delightful even for the mathematically challenged — are John L. Casti and Werner DePauli's *Gödel: A Life of Logic* and Rebecca Goldstein's *Incompleteness: The Proof and Paradox of Kurt Gödel*. Still, these treatments — simplified and made remarkably palatable to the nonmathematician — require patience and fortitude. More challenging, but widely acclaimed for its clarity and accuracy and for its critique of popular invocation and misuses of Gödel, is the late Torkel Franzen's *Gödel's Theorem: An Incomplete Guide to Its Use and Abuse*, cited above.

9. Ray Monk, *Bertrand Russell: The Spirit of Solitude, 1872–1921* (New York: Free Press, 1996), 118.

10. Bertrand Russell, *My Philosophical Development* (New York: Routledge, 1995; first published in 1959), 57.

11. Bertrand Russell, *The Autobiography of Bertrand Russell* (London: Routledge, 1998), 150.

12. The *Principia* was destined to become a landmark of modern mathematics. Still, Cambridge University Press shied away from publishing on such a daunting subject, fearing a loss of revenue. Russell and Whitehead were forced to ante up fifty pounds each for publication costs. *Autobiography*, 155.

13. Rebecca Goldstein, *Incompleteness: The Proof and Paradox of Kurt Gödel* (New York: Norton, 2006), 111–113.

14. Monk, *Spirit of Solitude*, 153–54.

15. Ibid., 154.

16. Dawson, 72.

17. Ibid., 77.

18. Bertrand Russell, *Problems of Philosophy* (Oxford: Oxford University Press, 1997; first published 1912), 3.

19. On Planck and black-body radiation, see Helge Kragh, *Quantum Generations: A History of Physics in the Twentieth Century* (Princeton: Princeton University Press, 1999), 58–64.

20. David Lindlay, *The End of Physics: The Myth of a Unified Theory* (New York: Basic Books, 1993), 11.

21. *Republic*, trans. Cornford, 1941, 527 (Stephanus numbers used).

22. *Insights of Genius* (Cambridge, MA: MIT Press, 2000), 180–82.

23. See Thomas S. Kuhn's *Black-Body Theory and the Quantum Discontinuity 1984–1912* (New York: Oxford University Press, 1978). Referenced in David C. Cassidy, *Einstein and Our World* (New York: Humanity Books, 2004), 53–54.

24. Albert Einstein, "Concerning an Heuristic Point of View Toward the Emission and Transformation of Light," *Ann. Phys.* 17, 132, 1905; Translation into English, *American Journal of Physics*, vol. 33, no. 5, May 1965.

25. Throughout this section, I am indebted to the following: John Gribbin, *Q Is for Quantum: An Encyclopedia of Particle Physics* (New York: The Free Press, 1998); J. P. McEvoy, *Introducing Quantum Theory* (Cambridge, UK: Icon Books, 1999); Tony Rothman, *Instant Physics* (New

York: Fawcett, 1995); David C. Cassidy, *Einstein and our World* (New York: Humanity Books, 2004); Michio Kaku, *Einstein's Cosmos* (New York: Atlas Books, 2004); Helge Kragh, *Quantum Generations: A History of Physics in the Twentieth Century* (Princeton: Princeton University Press, 1999); and Richard P. Feynman, *Six Easy Pieces* (New York: Basic Books, 1995; first published in 1963).

26. See "J. J. and the Cavendish," by Sir G. P. Thomson, at History of the Department, Department of Physics, University of Cambridge, at http://www.phy.cam.ac.uk/cavendish/history/years/jjandcav.php.

27. De Broglie's Nobel speech is quoted in the Mactutor History of Mathematics biography of de Broglie by J. J. O'Connor and E. F. Robertson, http://www-gap.dcs.st-and.ac.uk/~history/Biographies/Broglie.html.

28. At Heisenberg's doctoral oral examination were his adviser, Arthur Sommerfeld, and Wilhelm Wein, an experimental physicist whose lab course Heisenberg took with ill-concealed disdain. So impoverished was Heisenberg's knowledge of the experimental side he could not answer Wein's question about a simple storage battery. The incensed Wein wanted to fail Heisenberg. Only Sommerfeld's support salvaged a pass — with the grade of III, to Pauli's grade of I, a summa cum laude. Humiliated, Heisenberg set off for Max Born's laboratory wondering whether the job offer still stood. It did. Born himself was more theorist than experimentalist. See David C. Cassidy, *Uncertainty: The Life and Science of Werner Heisenberg* (New York: W. H. Freeman, 1992), 151–53.

29. Werner Heisenberg, *Physics and Beyond* (New York: Harper & Row, 1971), 38.

30. See "The Double-Slit Experiment" and "The Most Beautiful Experiment," *Physics Web*, September 2003. Young's experiment actually ranked fifth in the contest; its application to electrons came in first. http://physicsweb.org/articles/world/15/9/2002.

31. Richard P. Feynman, *Six Easy Pieces* (New York: Basic Books, 1995), 132.

32. Niels Bohr, "The Bohr-Einstein Dialogue," in *Niels Bohr: A Centenary Volume*, ed. A. P. French and P. J. Kennedy (Cambridge: Harvard University Press, 1985), 124.

33. Fölsing, 693.

34. Michio Kaku, *Einstein's Cosmos* (New York: W.W. Norton, 2004), 176.

35. Kaku calls these particles a "motley collection," compiled over the course of nearly 150 years, from the discovery of the cathode ray, which turned out to be the electron, to the "tau neutrino," discovered in 2000. Every few years, it seemed, another particle found its way into Greek nomenclature, to perhaps some consternation. Kaku quotes Oppenheimer: "The Nobel Prize in Physics should be given to the physicist who does *not* discover a new particle that year" (Kaku, 225).

36. *The Born-Einstein Letters,* 88.

37. Ibid., 146.

38. Ibid., 152, 161, 163.

39. Ibid., 165, 170, 171–72.

40. Ibid., 216.

41. Kaku, 232.

42. Etienne Klein and Marc Lachieze-Rey, *The Quest for Unity,* trans. Axel Reisinger (New York: Oxford University Press, 1999), 41.

43. Abraham Pais, *Subtle Is the Lord* (Oxford: Oxford University Press, 1982), 235.

44. "Autobiographical Notes," in *Albert Einstein: Philosopher Scientist, The Library of Living Philosophers,* ed. Paul Arthur Schilpp (New York: Tutor Publishing, 1949), 88–89.

45. Pais, 152.

46. *Einstein-Besso Correspondence,* 138. Translation by Burton Feldman and Katherine Williams.

47. Fölsing, 556.

48. *Ideas and Opinions,* 274.

49. Ibid., 233.

50. Quoted in Fölsing, 561, from a letter to Cornelius Lanczos dated January 24, 1938. Ernst Mach (1838–1916) was an influential and rigorous empiricist whose influence Einstein always acknowledged.

51. *Einstein and the History of General Relativity,* ed. Don Howard and John Stachel (Boston: Birkhauser, 1989), 315.

52. Note that in *The Evolution of Physics,* 257–58, Einstein commented on the relationship between matter and the energy of the field.

53. Pais, 141.

54. *Ideas and Opinions,* 230 (italics added).

55. Max Born and Albert Einstein, *The Born-Einstein Letters,* trans. Irene Born (New York: Macmillan, 2005), 82.

56. Ibid., 85.

57. Klein, 70.

58. Greene, 12.

59. Burton Feldman, *The Nobel Prize,* (New York: Arcade, 2000), 164.

60. Klein, 116.

61. Pais, 343.

62. Ibid., 343–50.

63. Ibid., 347.

64. Ibid., 350.

65. Enz, ed., *Pauli: Writings,* 116.

66. See Stanley Jaki, *A Mind's Matter* (Grand Rapids, MI: Wm. B. Eerdmans, 2002), 1–5.

67. Harald Atmanspacher and Hans Primas, "Pauli's Ideas on Mind and Matter in the Context of Contemporary Science" in *The Journal of Consciousness Studies,* 13, 3, 5–50, 2006.

68. Russell, *Human Knowledge,* 422.

69. "Remarks on Bertrand Russell's Theory of Knowledge," reprinted in *Ideas and Opinions,* 21.

70. Ibid., 24–25.

71. *Collected Papers,* vol. 11, 30.

72. Fölsing, 559. To Fölsing, Einstein's search for a unified theory was tainted and doomed by the belief that "mathematical criteria were 'the only reliable source of truth.'" See 561 and 559ff.

73. Russell, "Einstein and the Theory of Relativity," *Collected Papers,* vol. 11, 581–82.

74. From *Scientific American,* April 1950.

PART 4

1. Paul Lawrence Rose, *Heisenberg and the Nazi Atomic Bomb Project* (Berkeley: University of California Press, 1998), 284.

2. Thomas Powers, *Heisenberg's War* (Cambridge, MA: Da Capo Press, 2000), 344.

3. David Cassidy, *Uncertainty: The Life and Science of Werner Heisenberg* (New York: W. H. Freeman, 1992), 30.

4. Ibid., 85.

5. Werner Heisenberg, *Physics and Beyond,* trans. Arnold J. Pomerans (New York: Harper & Row, 1972), 93.

6. Mark Walker, *Nazi Science: Myth, Truth and the German Atomic Bomb* (New York: Plenum Press, 1995), 77.

7. Ibid., 83.

8. Ibid., 84.

9. Rainer Karlsch, *Hitlers Bombe* (München: Deutsche Verlas-Anstalt, 2005).

10. Rose, 260.

11. Ibid., 269.

12. Ibid., 238.

13. Ibid., 239.

14. Ibid., 248, 258.

15. Walker, 230.

16. Peter Goodchild, *J. Robert Oppenheimer: Shatter of Worlds* (New York: Fromm Intl., 1985), 61.

17. Jane S. Wilson and Charlotte Serber, *Standing By and Making Do* (Los Alamos, NM: Los Alamos Historical Society, 1988), 4.

18. Richard Rhodes, *The Making of the Atomic Bomb* (New York: Simon & Schuster, 1988), 523.

19. Ibid., 524.

20. Ibid.

21. S.S. Schweber, *In the Shadow of the Bomb* (Princeton, NJ: Princeton University Press, 2000), 70.

22. Rhodes, 444.

23. Ibid., 445.

24. Ibid., 449.

25. Ibid., 605.

26. Ibid., 571.

27. Paul Johnson, *Modern Times* (New York: Harper Collins, 2001), 205.

28. Schweber, 110.

29. Ibid., 123–24.

30. Ibid., 127.

31. Peter Michelmore, *The Swift Years: The Robert Oppenheimer Story* (New York: Dodd, Mead, 1969), 223.

32. Fred Jerome, *The Einstein File* (New York: St. Martins, 2002), 5–6.

33. Ibid., 39–40.

34. Schweber, 17.

EPILOGUE

1. Abraham Pais, *Subtle Is the Lord* (Oxford: Oxford University Press, 1982), 467.

2. Gerald Holton and Yehuda Elkaka, eds., *Albert Einstein: Historical Cultural Perspectives: The Centennial Symposium in Jerusalem* (Princeton: Princeton University Press, 1982), 398.

3. Pais, 341.

4. John A. Wheeler, *Geons, Black Holes, and Quantum Foam: A Life in Physics* (New York: Norton, 1998), 237–38.

5. Pais, 14–15.

6. David Brewster, *Memoirs of the Life, Writings, and Discoveries of Sir Isaac Newton* (New York: Johnson Reprint Corp., 1965), 143.

7. Burton Feldman, *The Nobel Prize* (New York: Arcade, 2000), 140.

8. S. S. Schweber, *In the Shadow of the Bomb* (Princeton, NJ: Princeton University Press, 2000), 65.

9. David Lindley, *The End of Physics: The Myth of a Unified Theory* (New York: Basic Books, 1993), 11. In Brian Greene's aptly titled *The Elegant Universe* (London: Vintage, 2005), the war between the experimentalists and theorists is dismissed in favor of the as-yet unproven string theories: "String theorists have no desire for a solo trek to the upper reaches of Mount Nature; the would far prefer to share the burden and the excitement with experimental colleagues. It is merely a technological mismatch in our current situation — a historical asynchrony — that the theoretical ropes and crampons for the final push to the top have at least been partially fashioned, while the experimental ones do not yet exist. But this does *not* mean that string theory is fundamentally divorced from experiment. Rather, string theorists have high hopes of 'kicking down a *theoretical* stone' from the ultra-high-energy mountaintop to experimentalists working at a lower base camp" (210). Let it not be said that physics is a classless society.

10. See, for instance, Michio Kaku's *Einstein's Cosmos* (New York: W. W. Norton, 2004), 230–33.

11. Charles P. Enz, *No Time to Be Brief: A Scientific Biography of Wolfgang Pauli* (Oxford: Oxford University Press, 2002), 389.

ACKNOWLEDGMENTS

THIS IS NOT THE BOOK Burton Feldman would have written. He died in early 2003, just after finishing a first draft. That, along with extensive notes, is what he left us. His wife and my dear friend, Peggy, believed with all her heart that the book should be published. She was too ill to take on the task of completing it, and so she turned to me. Nothing would have given her more joy than to have seen this book in print.

I first met Burton and Peggy as a young teenager. They became my lifelong friends. For a time, Burton was also my teacher at the University of Denver. I have fond memories of Burton in the classroom. Never have I met so natural a teacher — undidactic, passionate, humorous, sometimes contentious, and always able to listen. His brilliance never excluded, never belittled, never competed. On the contrary, it was impossible to feel anything but intelligent and worthy when in Burton's presence. With this book I feel blessed, as if Burton had given me yet another gift, one through which I could imagine the two of us in conversation again.

No one knew better than Burton how profoundly any piece of writing changes through the arduous and essential process of revision. The book's central premises and core research are his alone. Only Burton, a polymath with an abiding faith in history, could have imagined so much from the scantily recorded, unrecoverable conversations of four men whom chance and World War II delivered to the insular town of Princeton. I have conjectured and completed where it seemed necessary. For any errors or omissions, I am entirely responsible.

The list of acknowledgments must be incomplete. Certainly, Burton consulted with colleagues and resources unknown to me. I have tried to include within the bibliography every possible source referenced in the research notes, as well as those I have used in completing the book. Burton would have wished to thank the staff of Penrose Library at the University of Denver. Robert Richardson, Nancy Hightower, Helene Orr, Maria "Mimi" Katzenback, Gerald Chapman, Tug Yourgrau, and David Markson were friends and colleagues to whom Burton turned for inspiration and critical acumen. Burton's sister, Eleanor Feldman Werlin, was a source of love and support. Dr. Maureen Onat and Mary Ann Coats cared for Peggy and became part of Burton's extended family in the last years of his life. Esther Oliveri carefully prepared the initial bibliography from Burton's library. Tad Spencer did much early editing and assembling of material. Elizabeth Richardson and Tad were there for Burton and Peggy and have also been there for me. I am grateful to them both.

I am indebted to friends and colleagues who have been generous with moral support and with advice in matters scientific and editorial, especially Ellen Katz, John Fitzgerald, Diane Marks, Paulette Toth, Zulema Seligsohn, and Thea Stone. Family, friends, and colleagues at New York Institute of Technology bore my frequent inattentiveness with grace and patience. I owe much to the library staff of NYIT for their 24/7 reference portal and to the New York Public Library for use of the Wertheim Study. Darcy Falkenhagen was a stalwart believer at an early, critical stage. James Jayo's steady hand helped shepherd the book through production. Above all, I thank Richard Seaver for his thoughtful editing, generosity, and guidance. His affection for Burton and Peggy has sustained this project in ways beyond measure.

Katherine Williams

INDEX

absolute idealism, 68
Adler, Friedrich, 30
aging, scientific discovery and, ix–x, 7–16, 193–195
alchemy, 7
analytic philosophy, xii, 71, 118
"Annus mirabilis" papers, 31–33
antimatter, 117
anti-Semitism, 99, 108–109, 118, 121, 170, 212
"Appeal to Europeans, An" (Nicolai), 38, 39, 52
"Appeal to the Cultured World" (Fulda), 38
Aristotle, 128, 135
arts, 11–12
Atom and Archetype (Pauli and Jung), 109
atomic bomb, x, 16–17, 171
    effect of, on science, 188–190
    Germany's quest for, 165–167, 171–174

Los Alamos project, xv, 16–17, 174–180, 183–185, 188, 189
atomic structure, 32–33, 135–138
Autobiography (Russell), 60–61, 76

Bach, Johann Sebastian, 11
Barnes, Albert, 78
Beck, Otto, 38
Beethoven, Ludwig van, 11
Ben-Gurion, David, 46
Bergmann, Gustav, 87
Berlin, 36–37, 42–43, 50, 165–167
Bertram, Franca, 110
Besso, Michele, 24–25, 30, 150–151
beta-decay, xvii, 106–107
Bethe, Hans, 95, 98, 185
Black, Dora, 64–65, 65–66
black-body radiation, xvii, 132–134
Bohr, Niels, xiv, 13–16, 93, 97, 100–106, 128, 137–139, 167